SpringerBriefs in Earth Sciences

For further volumes:
http://www.springer.com/series/8897

Zhongliang Wu · Changsheng Jiang · Xiaojun Li
Guangjun Li · Zhifeng Ding

Earthquake Phenomenology from the Field

The April 20, 2013, Lushan Earthquake

Springer

Zhongliang Wu
Changsheng Jiang
Xiaojun Li
Zhifeng Ding
China Earthquake Administration
Institute of Geophysics
Beijing
China

Guangjun Li
Earthquake Administration of Sichuan Province
Chengdu
China

ISSN 2191-5369 ISSN 2191-5377 (electronic)
ISBN 978-981-4585-13-2 ISBN 978-981-4585-15-6 (eBook)
DOI 10.1007/978-981-4585-15-6
Springer Singapore Heidelberg New York Dordrecht London

Library of Congress Control Number: 2014939941

Printed on acid-free paper

Springer is part of Springer Science+Business Media (www.springer.com)

Dedicated to Jiayong He (1962–2014)
one of the active team members of the Field
Investigation of the 2013 Lushan Earthquake

Contents

Contributors

Zhifeng Ding Institute of Geophysics, China Earthquake Administration, Beijing, China

Lihua Fang Institute of Geophysics, China Earthquake Administration, Beijing, China

Shuang Gao Institute of Geophysics, China Earthquake Administration, Beijing, China

Changsheng Jiang Institute of Geophysics, China Earthquake Administration, Beijing, China

Hui Jiang Institute of Geophysics, China Earthquake Administration, Beijing, China

Guangjun Li Earthquake Administration of Sichuan Province, Chengdu, China

Ming Li China Earthquake Administration, Beijing, China

Xiaojun Li Institute of Geophysics, China Earthquake Administration, Beijing, China

Yingchun Li Earthquake Administration of Jiangsu Province, Nanjing, China

Aiwen Liu Institute of Geophysics, China Earthquake Administration, Beijing, China

Tengfei Ma Institute of Geophysics, China Earthquake Administration, Beijing, China

Zehong Mei Institute of Geophysics, China Earthquake Administration, Beijing, China

Xingchen Wang Institute of Geophysics, China Earthquake Administration, Beijing, China

Yushi Wang Institute of Geophysics, China Earthquake Administration, Beijing, China

Zhongliang Wu Institute of Geophysics, China Earthquake Administration, Beijing, China

Shengfeng Zhang Institute of Geophysics, China Earthquake Administration, Beijing, China

Chapter 1
The April 20, 2013, Lushan, Sichuan, China, Earthquake: An Overview

Abstract The April 20, 2013, Lushan, Sichuan, China, earthquake is one of the significant earthquakes in China in the 21st century. The earthquake was a test of the enhancement of the capacity to protect against earthquake disasters since the 2008 Wenchuan earthquake, and a test of earthquake science and technology in China.

Keywords The Lushan earthquake · Source parameters · Earthquake disasters

1.1 Parameters of the Earthquake

The Lushan earthquake occurred on April 20, 2013, at 08:02:46.0 local time, with epicenter location 30.3°N × 103.0°E, with focal depth 13 km, located near the Lushan County, Ya'an City, Sichuan Province (official data from the China Earthquake Administration on April 20, 2013). Figure 1.1 gives the location of the Lushan earthquake which ruptured a part of the southern Longmenshan fault zone and its relation with the 2008 Wenchuan earthquake which ruptured the middle and northern Longmenshan fault zone, in the context of the major to great earthquakes in China in the 20th century and the significant earthquakes in China since 2000 (Table 1.1).

Focal mechanism solutions from different seismological agencies provided a relatively consistent result of thrust-dominated rupture (Fig. 1.2). The rupture process from different seismological agencies showed a simple picture, with slip distributions relatively similar to each other, albeit different in details

This chapter is written by Zhang SF, Jiang CS, Jiang H, and Wu ZL (Institute of Geophysics, CEA).

Z. Wu et al., *Earthquake Phenomenology from the Field*, SpringerBriefs in Earth Sciences, DOI: 10.1007/978-981-4585-15-6_1,

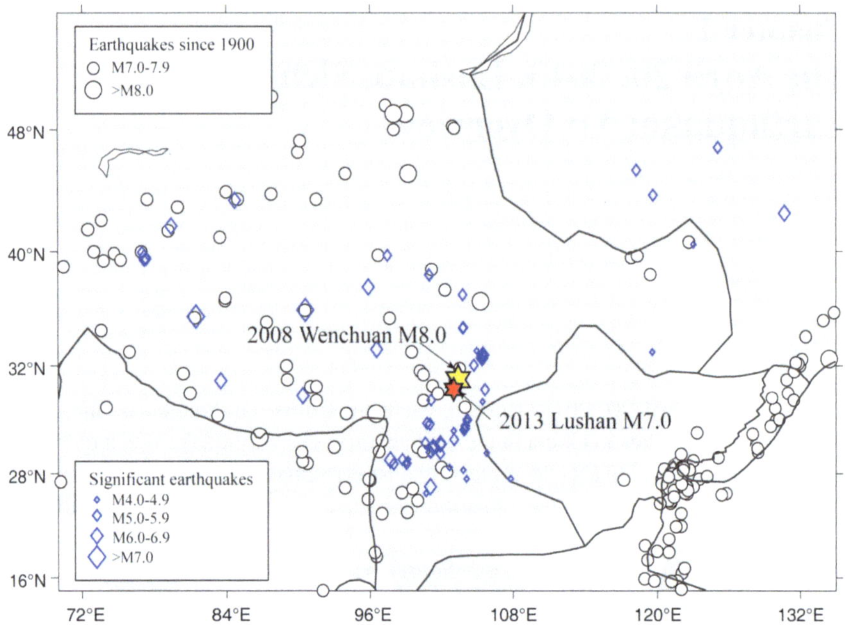

Fig. 1.1 Location of the Lushan earthquake, in the context of the seismicity in China. The *white circles* show the earthquakes in China and surrounding region with magnitude no less than M_W7.0 since 1900. Data from the IASPEI global catalogue 1900–1999 (Engdahl and Villaseñor 2002) and the GCMT catalogue since 2000. The *diamonds mark* the significant earthquakes in China since 2000 (Here significant earthquakes are defined as the earthquakes of magnitude 6.5 or greater, or ones that caused fatalities, injuries or substantial damage. From: http://earthquake.usgs.gov/earthquakes/eqarchives/significant/), data according to Table 1.1. Plate boundaries based on Bird (2003) are also shown in the figure

and in the centroid depth of the main asperity (Fig. 1.3). Even though with the differences which might be from the different seismological/geodetic/strong motion data used in the inversion and the Earth structures used to construct the Green's function, these results constrained the centroid depth of the main asperity within the range of 7–14 km. Moment magnitude M_W distributes from 6.5 to 6.8.

The National Earthquake Information Center (NEIC) of the US Geological Survey (USGS) obtained that the radiated energy from the earthquake[1] is $2.3 \pm 0.5 \times 10^{14}$ Nm, equivalent to the energy magnitude 6.7 and apparent stress 0.64 MPa.

Magnitude of the Lushan earthquake was determined by the China Earthquake Networks Center (CENC) of the China Earthquake Administration (CEA) as M_S7.0 and by the USGS/NEIC as M_S6.9. Figure 1.4 shows the station-based measurements

[1] http://earthquake.usgs.gov/earthquakes/eqarchives/energy/neic_b000gcdd_e.php

Table 1.1 Significant earthquakes in China in the 21st century

Date	Time UTC	Lat(°N)	Lon(°E)	M	Region	Comments
Jan 11, 2000	23:43:56.4	40.498	122.994	4.7	Northeastern China	Injured: 30
Jan 14, 2000	23:37:07.8	25.607	101.063	5.9	Yunnan	Killed: 7 Injured: 2528
Jan 26, 2000	20:55:19.1	24.263	103.797	4.5	Yunnan	Injured: 2
Jun 06, 2000	10:59:09.7	37.012	103.791	5.6	Gansu	Injured: 20+
Feb 23, 2001	00:09:23.6	29.513	101.129	5.6	Sichuan	Killed: 3 Injured: 109
Apr 12, 2001	10:47:00.3	24.768	99.061	5.6	Yunnan	Killed: 2+ Injured: 190
May 23, 2001	21:10:43.9	27.689	101.003	5.5	Sichuan	Killed: 2 Seriously Injured: 27 Slightly Injured: 578
Jun 07, 2001	18:03:32.3	24.785	99.038	4.6	Yunnan	Injured: 13+
Jul 14, 2001	18:36:08.3	24.455	102.660	4.6	Yunnan	Injured: 2
Oct 27, 2001	05:35:39.7	26.316	100.648	5.6	Yunnan	Killed: 1+ Injured: 220
Nov 14, 2001	09:26:10.0	35.946	90.541	7.8	Qinghai	
Dec 14, 2002	13:27:29.1	39.736	97.443	5.6	Gansu	Killed: 2
Feb 24, 2003	02:03:41.4	39.610	77.230	6.3	Southern Xinjiang	Killed: 261+ Injured: 4,000
Feb 25, 2003	03:52:41.0	39.483	77.393	5.4	Southern Xinjiang	Killed: 5
May 04, 2003	15:44:35.5	39.430	77.219	5.8	Southern Xinjiang	Killed: 1 Injured: 3
Jul 21, 2003	15:16:31.9	25.975	101.290	6.0	Yunnan	Killed: 16+ Injured: 584
Aug 16, 2003	10:58:42.7	43.770	119.643	5.4	Eastern Nei Mongol	Killed: 4+ Injured: 1000+
Oct 16, 2003	12:28:09.0	25.954	101.254	5.6	Yunnan	Killed: 3+ Injured: 32
Oct 25, 2003	12:41:35.2	38.400	100.951	5.8	Gansu-Qinghai border region	Killed: 9+ Injured: 43+
Oct 25, 2003	12:47:58.8	38.383	100.975	5.8	Gansu-Qinghai border region	
Nov 13, 2003	02:35:10.3	34.712	103.834	5.1	Gansu	Killed: 1+ Injured: 30
Nov 14, 2003	18:49:46.5	27.372	103.971	5.6	Sichuan-Yunnan-Guizhou region	Killed: 4+ Injured: 65
Nov 26, 2003	13:38:57.8	27.283	103.753	4.7	Sichuan-Yunnan-Guizhou region	Injured: 4
Mar 24, 2004	01:53:49.4	45.382	118.256	5.5	Eastern Nei Mongol	Injured: 100+
Aug 10, 2004	10:26:14.7	27.266	103.873	5.4	Sichuan-Yunnan-Guizhou region	Killed: 4+ Seriously Injured: 200 Slightly Injured: 400
Sep 07, 2004	12:15:49.7	34.682	103.781	5.2	Gansu	Injured: 9+
Oct 18, 2004	22:11:44.9	25.073	99.169	4.8	Yunnan	Injured: 12+
Jan 25, 2005	16:30:38.9	22.526	100.709	4.8	Yunnan	Injured: 3+
Feb 14, 2005	23:38:08.6	41.728	79.440	6.1	Southern Xinjiang	

(continued)

Table 1.1 continued

Date	Time UTC	Lat(°N)	Lon(°E)	M	Region	Comments
Jul 25, 2005	15:43:41.1	46.827	125.058	5.0	Heilongjiang	Killed: 1+ Injured: 12
Aug 05, 2005	14:14:48.0	26.569	103.036	5.2	Yunnan	Injured: 9+
Aug 13, 2005	04:58:44.8	23.627	104.103	4.8	Yunnan	Injured: 26+
Oct 27, 2005	11:18:57.3	23.604	107.798	4.2	Guangxi	Killed: 1 Injured: 1
Nov 08, 2005	07:54:38.9	9.973	108.287	5.3	South China Sea	Killed: 1
Jun 20, 2006	16:52:57.9	33.068	104.950	4.9	Gansu	Injured: 5
Jul 22, 2006	01:10:29.0	27.995	104.138	4.9	Sichuan-Yunnan-Guizhou region	Killed: 22 Injured: 106+
Aug 25, 2006	05:51:44.3	28.012	104.151	5.0	Eastern Sichuan	Killed: 1+ Injured: 31
Jun 02, 2007	21:34:57.7	23.028	101.052	6.1	Yunnan	Killed: 3 Injured: 329+
Mar 20, 2008	22:32:57.9	35.490	81.467	7.2	Xinjiang-Xizang border region	
May 12, 2008	06:28:01.5	31.002	103.322	7.9	Eastern Sichuan	Killed: 69,195 Injured: 374,177 Missing: 18,392
May 25, 2008	08:21:49.9	32.560	105.423	6.1	Sichuan-Gansu border region	Killed: 8 Injured: 927
May 27, 2008	08:37:51.5	32.710	105.540	5.7	Sichuan-Gansu border region	
Jun 17, 2008	05:51:43.2	32.761	105.554	4.8	Sichuan-Gansu border region	Killed: 2 Injured: 1
Jul 23, 2008	19:54:44.1	32.752	105.498	5.5	Sichuan-Gansu border region	
Jul 24, 2008	07: 09:30.0	32.747	105.542	5.7	Sichuan-Gansu border region	Killed: 1 Injured: 17
Aug 01, 2008	08:32:43.6	32.033	104.722	5.7	Sichuan-Gansu border region	Injured: 231+
Aug 05, 2008	09:49:17.2	32.756	105.494	6.0	Sichuan-Gansu border region	Killed: 4 Injured: 29
Aug 21, 2008	12:24:30.9	25.039	97.697	6.0	Myanmar- China border region	Killed: 5 Injured: 127
Aug 25, 2008	13:21:58.8	30.901	83.520	6.7	Western Xinjiang	
Aug 30, 2008	08:30:53.0	26.241	101.889	6.0	Sichuan-Yunnan border region	Killed: 43+ Injured: 585
Aug 31, 2008	08:31:10.7	26.232	101.970	5.6	Sichuan-Yunnan border region	Killed: 2
Oct 06, 2008	08:30:45.5	29.807	90.350	6.3	Eastern Xinjiang	Killed: 10 Injured: 25+
Nov 10, 2008	01:22:02.5	37.565	95.833	6.3	Northern Qinghai	Injured: 3
Dec 09, 2008	18:53:11.1	32.518	105.395	5.1	Sichuan-Gansu border region	Killed: 2+ Injured: 3
Jul 09, 2009	11:19:16.2	25.632	101.095	5.7	Yunnan	Killed: 1 Injured: 336
Aug 08, 2009	13:26:17.5	29.358	105.438	3.7	Sichuan-Chongqing border region	Killed: 2+ Injured: 1
Nov 01, 2009	21:07:20.6	25.962	100.825	4.9	Yunnan	Injured: 28

(continued)

Table 1.1 continued

Date	Time UTC	Lat(°N)	Lon(°E)	M	Region	Comments
Jan 17, 2010	09:37:26.1	25.558	105.804	4.4	Guizhou	Killed: 7+ Injured: 9 Missing: 1
Jan 30, 2010	21:36:58.0	30.268	105.668	5.1	Eastern Sichuan	Killed: 1 Injured: 15
Feb 18, 2010	01:13:19.5	42.587	130.703	6.9	China-Russia-North Korea border region	
Feb 25, 2010	04:56:51.9	25.523	101.903	5.2	Yunnan	Injured: 11
Apr 13, 2010	23:49:38.3	33.165	96.548	6.9	Southern Qinghai	Killed: 2,698+ Injured: 12,135 Missing: 270
Aug 29, 2010	00:53:31.4	27.197	103.005	4.9	Sichuan-Yunnan-Guizhou Region	Injured: 14
Mar 10, 2011	04:58:16.0	24.719	97.969	5.5	Myanmar- China border region	Killed: 25 Injured: 250
Jun 20, 2011	10:16:55.2	25.075	98.721	5.3	Myanmar- China border region	Injured: 4
Aug 11, 2011	10:06:29.3	39.955	77.028	5.6	Southern Xinjiang	Injured: 21
Jun 24, 2012	07:59:34.8	27.767	100.781	5.5	Sichuan-Yunnan border region	Killed: 4 Injured: 394+
Jun 29, 2012	21:07:33.8	43.433	84.700	6.3	Northern Xinjiang	Injured: 52+
Jul 20, 2012	12:11:52.0	32.978	119.593	4.9	Jiangsu	Killed: 1 Injured: 2
Sep 07, 2012	03:19:42.5	27.575	103.985	5.5	Sichuan-Yunnan-Guizhou region	Killed: 81+ Injured: 821

Data from USGS (http://earthquake.usgs.gov/earthquakes/eqarchives/significant/)

of the surface wave magnitude. Chinese M_S measures the peak amplitude on the simulated mid-to-long-period SK recordings (along the horizontal component), using[2]

$$M_S = \lg(\frac{A}{T})\text{max} + 1.66\log(\Delta) + 3.5 \quad (1.1)$$

Difference is that the measurement spans the distance range $3^o < \Delta < 177^o$ with periods $T > 6$ s, therefore it is more ‘broadband’ and ‘wide-range’. Statistically, the Chinese M_S is more close to the IASPEI recommended broadband surface wave magnitude, and is on average 0.2 higher than the NEIC-measured M_S (Liu et al. 2006). In this regard, the magnitude readings 7.0 (CENC) and 6.9 (NEIC) is in good agreement.

Mentioning this agreement is also important because this earthquake seems somehow ‘strange’ in that the surface wave magnitude M_S is apparently larger than, at least not less than, the moment magnitude M_W, which is different from the textbook knowledge of ‘magnitude saturation’. The first suspect was naturally

[2] Chinese Standard GB 17740-1999 General ruler for earthquake magnitude (in Chinese).

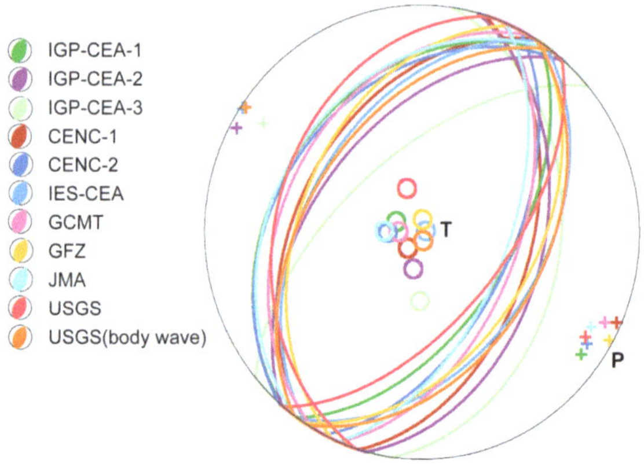

Fig. 1.2 Focal mechanism solutions of the Lushan earthquake from different agencies, as marked by different colors. *Circles* and *crosses* in the figure denote the T and P axes, respectively. Data from Liu et al. (2013a), in which most of the results from different groups are based on the online information (In the figure, IGP-CEA-1: contributed by Dr. C. Liu from the Institute of Geophysics, China Earthquake Administration (IGP-CEA) using teleseismic waveforms; IGP-CEA-2: contributed by Dr. L. B. Han from the IGP-CEA using regional waveforms; IGP-CEA-3: contributed by X. Y. Guo from the IGP-CEA using P-wave first motion; CENC-1: contributed by Prof. R. F. Liu from the CENC using teleseismic waveforms; CENC-2: contributed by Dr. X. Zhao from the CENC using teleseismic waveforms; IES-CEA: contributed by Dr. Q. C. Wang from the Institute of Earthquake Science, China Earthquake Administration (IES-CEA) using regional waveforms.)

that the surface wave magnitude or the moment magnitude may have some problems, which led to the careful re-check of the recordings (Liu et al. 2013a).

1.2 Disasters of the Earthquake and Timeliness of the Emergency Response

1.2.1 Emergency Response to the Earthquake

Emergency response to the Lushan earthquake can be seen from the reports of news media such as SinaNews.[3] Following is a selected list of important events based on SinaNews. Note that in the list, the time of each piece of news is the time of publishing the report, which is later than the event being reported.

- *April 20, reported 08:13* The Lushan M_S7.0 earthquake occurred
- *April 20, reported 08:56* Sichuan Province Earthquake Administration invoked the earthquake emergency response action

[3] http://roll.news.sina.com.cn/s_yadzh2013_all/index.shtml (in Chinese).

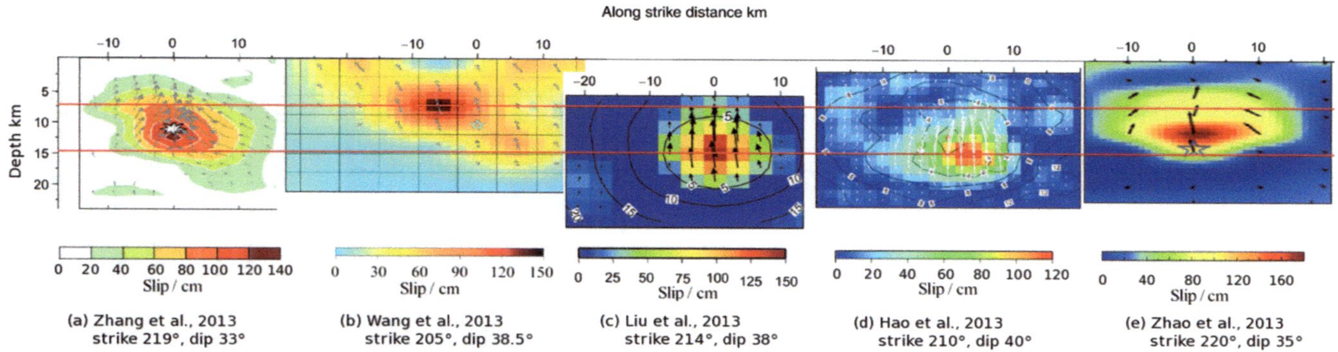

Fig. 1.3 Slip distribution from different Chinese seismological agencies, projected to the vertical direction (along depth). From *left* to *right* the Institute of Geophysics, China Earthquake Administration (Zhang et al. 2013); the Institute of Tibetan Plateau Research, Chinese Academy of Sciences (Wang et al. 2013); the Institute of Geodesy and Geophysics, Chinese Academy of Sciences (Liu et al. 2013b); the Institute of Geology and Geophysics, Chinese Academy of Sciences (Hao et al. 2013); and the Institute of Earthquake Science, China Earthquake Administration (Zhao et al. 2013). Arbitrarily ordered, without a unified color bar of static slip. The two horizontal *solid lines* in *red* labeled the upper and lower bound of the centroid depth of the main asperity provided by different agencies

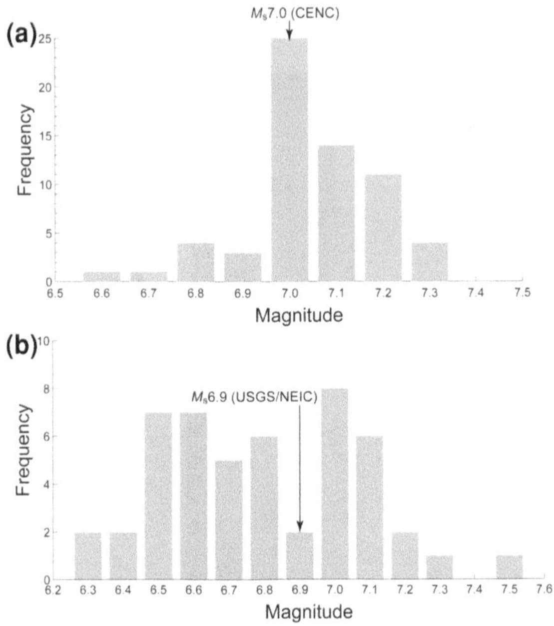

Fig. 1.4 Distribution of station-based magnitude measurements: **a** Chinese stations, measured by the CENC; **b** worldwide stations, measured by the NEIC. Data from Liu et al. (2013a)

Fig. 1.5 A bbs picture showing the scene of on-site rescue

- *April 20, reported 09:36* Emergency and rescue team of Sichuan Province arrived at Ya'an and started the rescue action
- *April 20, reported 09:36* Sichuan Province invoked the Class I disaster relief response action
- *April 20, reported 10:11* The Chengdu Area Command of the Chinese People's Liberation Army (PLA) sent 2000 troops to Lushan (Fig. 1.5 is a bbs picture showing the scene of the rescue)
- *April 20, reported 10:33* Central government invoked the Class III disaster relief response action
- *April 20, reported 10:33* China Earthquake Administration (CEA) invoked Class I earthquake emergency response action

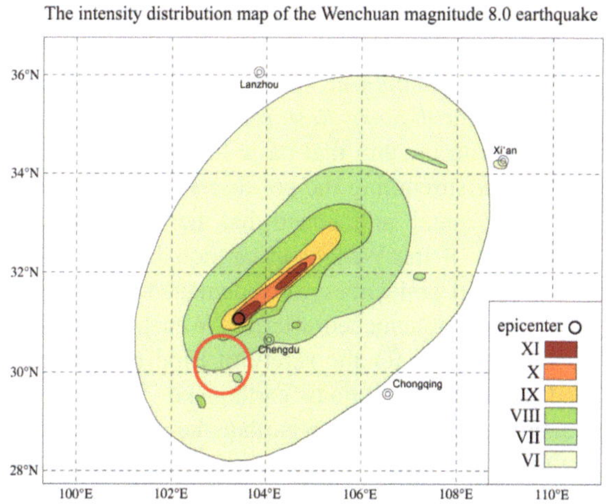

Fig. 1.6 Location of the Lushan earthquake (the *red circle*), in the context of the intensity distribution of the Wenchuan earthquake (From: http://www.cea.gov.cn/manage/html/8a8587881632fa5c0116674a018300cf/_content/08_08/29/1219980517676.html) which is officially released by the China Earthquake Administration (CEA), translated from Chinese

- *April 20, reported 11:09* Medical team arrived at Lushan
- *April 20, reported 11:44* State Council invoked Class I earthquake emergency response action
- *April 20, reported 12:32* Premier Li Keqiang flew to the epicentral area
- *April 21, reported 02:35* Disaster relief supporting materials arrived at the disastrous region
- *April 21, reported 17:53* Electric power supply restored in the disastrous region
- *April 23, reported 09:09* Chinese President Xi Jinping expressed via telephone the sympathy to the people in the disastrous region
- *April 23, reported 17:04* Road transportation and postal service in all Sichuan restored
- *April 25, reported 17:45* China Earthquake Administration (CEA) completed the intensity map
- *April 26, reported 02:06* Sichuan Province government decided that April 27 be the provincial mourning day for the earthquake victims.

According to the China Earthquake Administration (CEA), the earthquake emergency stage ended on May 5, 2013. On July 15, 2013, the comprehensive plan for the post-disaster reconstruction was ratified.

1.2.2 Disasters of the Lushan Earthquake Testing the Wenchuan Reconstruction

The 2013 Lushan earthquake occurred in the intensity VII region of the 2008 Wenchuan earthquake (Fig. 1.6). The Wenchuan earthquake (see Chen and Booth 2011, for a systematic review; see Wu and Ma 2014, for a concise introduction), with maximum intensity XI, affected 417 county-level units (county, county-level-city, or district, according to the terminology of the Chinese governmental system) of 10 provinces, among which there were 51 county-level units hard-hit and/or extremely hard-hit, mainly in Sichuan, Gansu, and Shaanxi Provinces, with a population about 20 million. On September 19, 2008, *The Overall Planning for Post-Wenchuan Earthquake Restoration and Reconstruction* was ratified by the State Council, with the aim that basic living conditions and the economic development reaching or surpassing the pre-disaster level. In mid-2011, after three years work, the State Council announced that the reconstruction has been completed. Taking the lessons of the Wenchuan earthquake that 7,444 schools and about 100 million m^2 rural residences were damaged, a national project of the reinforcement of schools was launched after the Wenchuan earthquake, and the pre-earthquake-launched project for the safety of rural residences was accelerated and strengthened after the Wenchuan earthquake. This endeavor significantly reduced the damage during the Lushan earthquake.

Officially released data on April 25 by the State Council announced that, up to 6:00 p.m., April 24, there had been 196 deaths plus 21 missing, and 13,484 injured.[4] Li et al. (2013) compared this earthquake with the 2010 Yushu M_S7.1 earthquake which caused more fatalities (2,698 deaths plus 270 missing) in a less populated area in Qinghai Province, and argued that this difference reflects the advancement obtained with the reconstruction after the Wenchuan earthquake. Despite their comparison apparently neglected the characteristics of the Yushu earthquake that the strike-slip earthquake rupture propagated (maybe with a super-shear speed) towards the Jiegu Town, Yushu, which had caused tremendous loss, it is true that the reconstructed buildings in Lushan, even if with damage caused by the earthquake-generated ground motion, played an important role in saving lives.

1.3 Studies on the Earthquake

Shortly after the earthquake, there were tens of research papers published related to the properties of the mainshock, the co-seismic variation, the seismic and geological disasters, the aftershock sequence, the geology and geophysics of the earthquake, the relation between the Lushan earthquake and the Wenchuan earthquake, and the impact of the Lushan earthquake on regional seismic hazard, as well as some retrospective case studies related to the precursor-like anomalies. Appendix A1 summarizes the publications on the Lushan earthquake. Due to the on-going nature of the related studies, this summary is still far from complete. However, it reflects the recent development of earthquake science and technology in China in a special perspective.

To understand the scientific problems associated with the Lushan earthquake, namely the seismogenic fault of the Lushan earthquake, the relation between the Lushan earthquake and the Wenchuan earthquake, and the time-dependent

[4] http://www.scio.gov.cn/xwfbh/xwbfbh/wqfbh/2013/0425p/

seismic hazard of the southern Longmenshan fault zone as well as its surrounding regions, a Field Investigation was organized by the CEA (see Chap. 2 of this Brief). Meanwhile, the Department for Earthquake Monitoring and Prediction, CEA, organized in parallel the Scientific Summary and Reflection on the Forecast of the Lushan Earthquake, a systematic retrospective case study both in a technical perspective and in a managing perspective, aiming at the accumulation of lessons and experiences for the study of earthquake predictability.

1.4 Summary

Comparing to other earthquakes in China, the Lushan earthquake is neither characterized by its large size nor featured by its tremendous disasters. It is even not puzzling in seismological observation and interpretation: a consistent focal mechanism result of thrusting fault, a consistent rupture process result of simple 'isotropic' rupture propagation which constrained the centroid depth within the range 7–14 km, and a consistent result of magnitude M_W6.5–6.8 and M_S6.9 (USGS/NEIC)–7.0 (CEA/CENC). To much extent, it is only 'one of the significant' earthquakes in China.

On the other hand, however, each earthquake has its own stories. The Lushan earthquake is of no exception. In this chapter, at least one of the characteristic features of this earthquake emerges, that comparing to other earthquakes in China, this earthquake might be the first one to have so many research results in such a short time. And some of the consistent result itself is a scientific problem, as per why the magnitude saturation did not appear for this event, that is, M_S is apparently larger than M_W. Moreover, comparing the Lushan earthquake with its nearby Wenchuan earthquake, several scientific questions could be raised, which are related not only to earthquake science but also to social sustainability.

References

Bird P (2003) An updated digital model of plate boundaries. Geochem Geophys Geosyst 4(3):1027. doi:10.1029/2001GC000252

Chen Y, Booth DC (2011) The Wenchuan earthquake of 2008: anatomy of a disaster. Science Press in Cooperation with Springer, Beijing

Engdahl ER, Villaseñor A (2002) Global seismicity: 1900–1999. In: Lee WHK, Kanamori H, Jennings PC, Kisslinger C (eds) International handbook of earthquake and engineering seismology, Academic Press, San Diego, Part A, pp 665–690

Hao JL, Ji C, Wang WM, Yao ZX (2013) Rupture history of the 2013 M_W 6.6 Lushan earthquake constrained with local strong motion and teleseismic body and surface waves. J Geophys Res 40:5371–5376. doi:10.1002/2013GL056876

Li ZQ, Hou JS, Li Y, Su GW, Jiang LX, Sun BT, Yuan YF, Lin JQ (2013) Analysis on the characteristics of the M_S7.0 Lushan, Sichuan Province, earthquake hazard on April 20, 2013. Seismol Geol 35:398–410. doi:10.3969/j.issn.0253-4967.2013.02.018 (in Chinese with English abstract)

Liu CL, Zheng Y, Ge C, Xiong X, Xu HZ (2013a) Rupture process of the *M*7.0 Lushan earthquake, 2013. Sci China Earth Sci 56:1187–1192. doi:10.1007/s11430-013-4639-9

Liu RF, Chen YT, Bormann P, Ren X, Hou JM, Zou LY, Yang H (2006) Comparison between earthquake magnitudes determined by China seismograph network and US seismograph network (II): surface wave magnitude. Acta Seismol Sin 19:1–7. doi:10.1007/s11589-001-0001-y

Liu RF, Chen YT, Zou LY, Chen HF, Liang JH, Zhang LW, Han XJ, Ren X, Sun L (2013b) Determination of parameters for the 20 April 2013 Lushan $M_W6.7(M_S7.0)$ earthquake in Sichuan Province. Acta Seismol Sin 35:652–660. doi:10.3969/j.issn.0253-3782.2013.05.004 (in Chinese with English abstract)

Wang WM, Hao JL, Yao ZX (2013) Preliminary result for rupture process of Apr. 20, 2013, Lushan earthquake, Sichuan China. Chin J Geophys 56:1412–1417. doi:10.6038/cjg20130436 (in Chinese with English abstract)

Wu ZL, Ma TF (2014) Chapter 23: The 2008 Wenchuan, China, earthquake. In: Ismail-Zadeh A, Urrutia-Fucugauchi J, Kijko A, Takeuchi K, Zaliapin I (eds) Extreme natural hazards, disaster risks and societal implications. Cambridge University Press, Cambridge, pp 301–309

Zhang Y, Xu LS, Chen YT (2013) Rupture of the Lushan 4.20 earthquake and preliminary analysis on the disaster-causing mechanism. Chin J Geophys 56:1408–1411. doi:10.6038/cjg20130435 (in Chinese with English abstract)

Zhao CP, Zhou LQ, Chen ZL (2013) Source rupture process of Lushan $M_S7.0$ earthquake, Sichuan, China and its tectonic implication. Chin Sci Bull 58:3444–3450. doi:10.1007/s11434-013-6017-6

Chapter 2
The Field Investigation of the Earthquake

Abstract Organized by the China Earthquake Administration (CEA), Field Investigation was conducted after the Lushan earthquake. The Field Investigation can be divided into two stages: the stage of emergency, and the post-emergency stage.

Keywords The field investigation · The earthquake emergency · The southern Longmenshan fault zone

2.1 Scientific Issues Related to the Field Investigation

In the context of the significant earthquakes in China (see Fig. 1.1), the April 20, 2013, Lushan earthquake occurred in the central-China north-south seismic zone, along the southern Longmenshan fault zone. The middle and north segment of the Longmenshan fault zone was ruptured by the 2008 Wenchuan earthquake. When speaking the Longmenshan fault 'zone', the four nearly parallelly-striking faults have to be accounted for, namely, from the west to the east, the Houshan fault (the back range fault, in Chinese Hou = back, Shan = mountain), the Zhongyang fault (the central fault, in Chinese Zhongyang = central), the Qianshan fault (the front range fault, in Chinese Qian = front, Shan = mountain), and the Shanqian fault (the range front fault). The first question about the Lushan earthquake, therefore, is which fault, or another unknown fault, is responsible for this earthquake.

This chapter is written by Wu ZL (Institute of Geophysics, CEA), Li GJ (Earthquake Administration of Sichuan Province), Li M (Department of Science and Technology, CEA), and Li XJ (Institute of Geophysics, CEA), serving as team leaders of the Field Investigation of the 2013 Lushan Earthquake. Materials are from the Field Investigation, with the helps from the Department for Earthquake Monitoring and Prediction, CEA; the Department for Emergency and Rescue, CEA; and the Administrative Office of CEA. Jiang H and Zhang SF helped in the data processing.

Z. Wu et al., *Earthquake Phenomenology from the Field*, SpringerBriefs in Earth Sciences, DOI: 10.1007/978-981-4585-15-6_2, © The Author(s) 2014

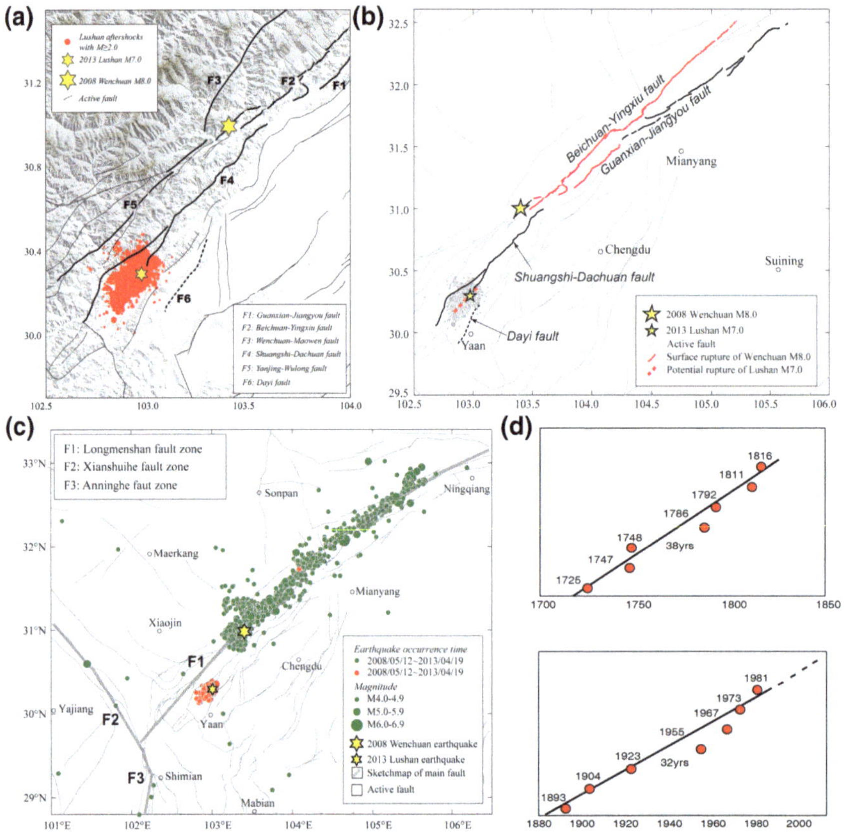

Fig. 2.1 **a** Active faults associated with the 2008 Wenchuan earthquake and the 2013 Lushan earthquake. *Yellow star* and *red star* show the epicenters of the Wenchuan earthquake and the Lushan earthquake, respectively. The aftershocks of the Lushan earthquake are also shown in the figure. Aftershock data from April 20 to June 14, 2013 with magnitude no less than M_L2.0, are from the CENC (http://10.5.202.22/bianmu/index.jsp). Active fault data is provided by the National Center for Active Fault Studies (NCAFS). **b** Epicenter and surface rupture of the 2008 Wenchuan earthquake, and deduced earthquake rupture in deep of the 2013 Lushan earthquake, together with the related faults. Data of surface fault of the Wenchuan earthquake is from Xu et al. (2008), in which the central fault is the Beichuan-Yingxiu fault, and the front range fault is the Guanxian-Jiangyou fault in the mid Longmenshan fault zone and the Shuangshi-Dachuan fault in the southern Longmenshan fault zone. **c** Aftershocks of the 2008 Wenchuan earthquake and the 2013 Lushan earthquake, together with the Xianshuihe fault zone, the Longmenshan fault zone, and the Anninghe fault zone, highlighting the ‘seismic gap’ between the Wenchuan aftershocks and the Lushan aftershocks, and between the Lushan aftershocks and the ‘triple junction’. **d** Strong earthquakes with magnitude larger than 6.5 along the Xianshuihe fault. *Top* the active period between 1725 and 1816; *Bottom* the active period since 1893

Figure 2.1a shows this situation. Answering this question is, to much extent, an interdisciplinary endeavor which needs to overcome the limitation of seismology by reducing the uncertainty of the of the earthquake location, to overcome the limitation of geology (which investigates the faults and paleo-earthquakes on a point-by-point basis, and sometimes cannot provide an accurate location of the faults) by organizing a comprehensive field investigation, and to overcome the limitation of geophysics by enhancing the resolution of geophysical mapping.

For the Lushan earthquake, the relation with the Wenchuan earthquake is another key issue to be considered. Figure 2.1b shows the locations of the Wenchuan earthquake sequence and the Lushan earthquake sequence. In between of the Lushan sequence and the Wenchuan sequence, the southern Longmenshan fault zone left a 45 km 'seismic gap'. Between the Lushan sequence and the 'triple junction' among the Xianshuihe, Anninghe, and Longmenshan fault zone there is another 'seismic gap' with comparable size (see Fig. 2.1c). The two 'seismic gaps' make the debate about the relation between the Lushan earthquake and the Wenchuan earthquake of practical implication to regional seismic hazard assessment. It has to be noticed that the Wenchuan earthquake was not a simple rupture. Rather it ruptured the central fault (in Fig. 2.1b the Beichuan-Yingxiu fault) and the front range fault (in Fig. 2.1b the Guanxian-Jiangyou fault), with surface ruptures 240 and 70 km, respectively. It was suspected that the Lushan earthquake occurred along the front range fault (in Fig. 2.1b the Shuangshi-Dachuan fault), or the Dayi fault, or 'another' unknown fault. These fault associations provide different models for both the seismogenesis of the Lushan earthquake and the seismic hazard in the southern Longmenshan fault zone.

Figure 2.1c shows the sketch of the two 'seismic gaps' along the southern Longmenshan fault zone. In the 'north gap' there was the record of a magnitude 6.2 earthquake in Dayi on February 24, 1970, approximately along the Dayi fault (as shown in Fig. 2.1a). In the 'south gap' there were the records of a magnitude 6 earthquake in Luding-Tianquan on June 12, 1941, and a magnitude 6 earthquake in Tianquan in September 1327. Problem is that the accurate locations of these two historical earthquakes could not be determined using historical recordings, with the uncertainty of up to 100–200 km.

The region of the 'triple junction' is even more complicated at the present time considering another famous fault, the Xianshuihe strike-slip fault. Figure 2.1d shows the strong earthquakes with magnitude larger than 6.5 along the Xianshuihe fault. Two active periods are shown in the figure, namely the active period between 1725 and 1816, and the active period since 1893. Note that till 2013 there has been 32 years since the last strong earthquake in 1981, in the active period since 1893 the maximum inter-seismic duration is 32 years, but the duration between the recent two active periods is 77 years, the seismic hazard along the Xianshuihe fault is not a simple picture.

Among the scientific problems associated with the Lushan earthquake, the fault association, the relation with the Wenchuan earthquake, and the seismic hazard of the southern Longmenshan fault zone are of special concern not only in earthquake science but also in the public. Responding to this concern, after the earthquake, organized by the China Earthquake Administration (CEA), the Field Investigation of the 2013 Lushan Earthquake was organized, participating in by the institutions in the CEA, with the assistance of the Earthquake Administration of Sichuan Province. Figure 2.2 shows the scientific targets of the

Fig. 2.2 Logo of the field investigation highlighting the scientific goals (Designed by Wu ZL, team leader of the field investigation of the 2013 Lushan earthquake.). In the logo the *black box* shows the rupture zone of the 2008 Wenchuan earthquake, and the *red circle* shows the aftershock zone of the 2013 Lushan earthquake. The *exclamation mark* consists of the Wenchuan earthquake and the Lushan earthquake, as shown in the logo, indicating a warning of the future seismic hazard

Field Investigation by its logo, which shows the Lushan earthquake, the Wenchuan earthquake rupture zone, and the Longmenshan fault zone. The yellow box is similar to the National Geographic, indicating the nature of the activity. The exclamation mark consisting of the Wenchuan earthquake and the Lushan earthquake indicates a warning of the future seismic hazard.

2.2 The Field Investigation of the 2013 Lushan Earthquake

2.2.1 Field Investigation in the Stage of Earthquake Emergency

The Field Investigation started almost at the same time when the earthquake stroke. As a matter of fact, institutions of the CEA had prepared a plan for the earthquake field investigation and the design of the organization. Generally, after an earthquake, the field investigation includes engineering and damage investigation, mobile seismic observation of aftershocks, seismo-tectonic field work, and geophysical explorations of the structure of seismic source. After many of the earthquakes, field investigation of the precursor-like anomalies is also organized.

Figure 2.3a shows the intensity distribution map officially released by the CEA in the emergency response stage. This map is to be revised and refined due to its emergency response nature. Figure 2.3b shows the working sites of the engineering group of the Field Investigation. Figure 2.3c shows structural engineers working in the disastrous region. These works are to much extent connected to the field work for earthquake emergency. In the CEA, the earthquake emergency includes two fronts: network-based seismological information service to the public and the government (see Chap. 6 of this Brief), and field work. The field work includes three components: rescue and relief actions, field investigation of damage and

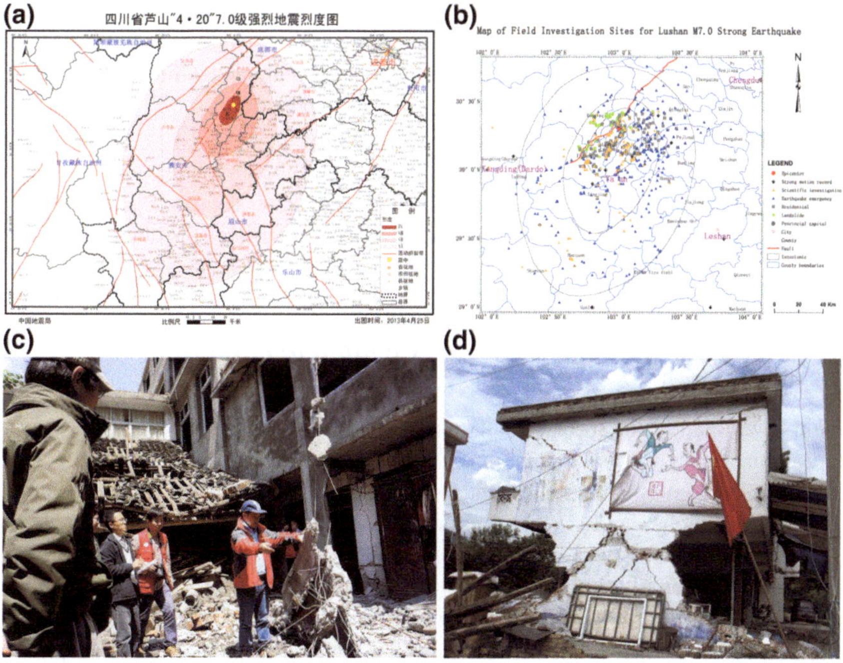

Fig. 2.3 **a** Intensity distribution of the Lushan earthquake, officially released by the CEA in the emergency response stage (From: http://www.cea.gov.cn/publish/dizhenj/464/478/20130425153642550719811/index.html), with the Chinese characters on the *top* 'The intensity distribution map of the April 20 Lushan, Sichuan Province magnitude 7.0 earthquake', subject to revision based on field investigation and study. See Fig. 2.3b. **b** Field investigation of engineering damages and disasters: Map used for the field investigation of the task force (unpublished, courtesy of Prof. Dai JW). *Blue triangle* sites of field investigation of engineering damage; *Black cross* strong ground motion recording stations; *Green dots* sites for remote sensing interpretation of landslides; *Gray squares* residential sites for remote sensing interpretation of damages. The epicenter (*red dot*), the Shuangshi-Dachuan fault (*red line*), and the (revised) isoseismal (*black thin line*) are also marked on the map. **c** Structural engineers in the field. Shown in the figure is Li XJ (the first on the *right*), associate team leader of the Field investigation of the 2013 Lushan earthquake, in the disastrous region in the days after the earthquake. Photo by the engineering group of the field investigation of the 2013 Lushan earthquake. **d** A typical 'standing ruin'. *Photo* by Guo X, Institute of Disaster Prevention

disaster assessment, and scientific works including mobile field observations of aftershocks and geophysical fields and field survey of seismo-tectonics.

It has to be noticed that field work has the unique role which cannot be replaced by remote sensing. Figure 2.3d shows a special kind of seismic destruction in the epicentral region, the 'standing ruin'. In the remote sensing picture, this ruin is hard to recognize if there is not special data processing added.

Before the Lushan earthquake, in connection to the post-Wenchuan earthquake field work, several continuous GPS stations were deployed in the Lushan region. These GPS stations were named as LS#, with # stands for numbers of different stations. Therefore some people thought that there must be some 'forecasts' before the Lushan earthquake. But actually LS is a Chinese abbreviation Lin-Shi (temporary) and has nothing to do with Lushan. And this misunderstanding just reflects a good hope of people to solve the problem of earthquake forecast.

Started from April 20, mobile seismic stations were deployed in the epicentral region, monitoring the aftershock activity. Up to April 25, 15 mobile seismic stations were deployed and networked. These near source observation stations, together with the local and regional seismic stations, well-constrained the aftershock distribution.

2.2.2 *Post-Earthquake Field Investigation and Studies*

Field investigation for surface faults and geological disasters were also organized in the emergency response stage, and lasted after the ending of the emergency. In the later field investigation, besides the systematic investigation of active faults, geological trenches and near-ground-surface geophysical exploration were conducted to search for the traces of the past earthquakes (Fig. 2.4), including the immediately passed earthquake (the Lushan earthquake). Field investigation obtained a 'zero result' that there was no surface rupture observed. What had been claimed to be the 'seismic surface faults' (e.g., Liu et al. 2014a) were all proved to be the secondary geological effects. It is worth mentioning that, like a detective searching and collecting evidences, or like Thomas Edison excluded the materials which could not be used in the electric lamp, such 'zero result' in the field investigation plays an equivalently important role for the scientific understanding of the earthquake as the discovered surface faults and historical earthquakes.

With the emergency response stage finalized, more mobile seismic stations were deployed and joined in the network. Seismologists from the Institute of Geophysics, CEA, and the Earthquake Administration of Sichuan Province learnt from IRIS the improvement of the site condition to enhance the quality of observation (Fig. 2.5a). By combing the mobile seismic stations, the reservoir seismic stations, the public/industry seismic stations, the local seismic stations operated by the Earthquake Administration of Sichuan Province, and the regional seismic stations operated by neighboring provinces such as Chongqing (Fig. 2.5b), the seismic network carried out a real-time monitoring of the Lushan sequence. Double difference (DD) relocation was conducted to the aftershock sequence (see Chap. 4 of this Brief). The aftershock sequence was also 'predictively monitored' by the ETAS model which estimates the expected earthquake rate.

Supported by, and in collaboration with, other existing projects, the Field Investigation of the 2013 Lushan earthquake also carried out geophysical mapping such as MT sounding, geomagnetic mapping, and gravity mapping, as well as leveling measurement for ground deformation. A wide-angle reflection profile, crossing the Longmenshan fault zone, was planned to map the structure of the seismic source. Such collaborative mechanism was also valid for the works of engineering damage investigation, seismo-tectonics and paleo-earthquakes, and seismological

Fig. 2.4 Geological trench to detect the earthquake fault, in Shuangshi Town, over the co-seismic liquefaction zone. *Picture* by the Seismo-tectonics group of the field investigation of the 2013 Lushan earthquake

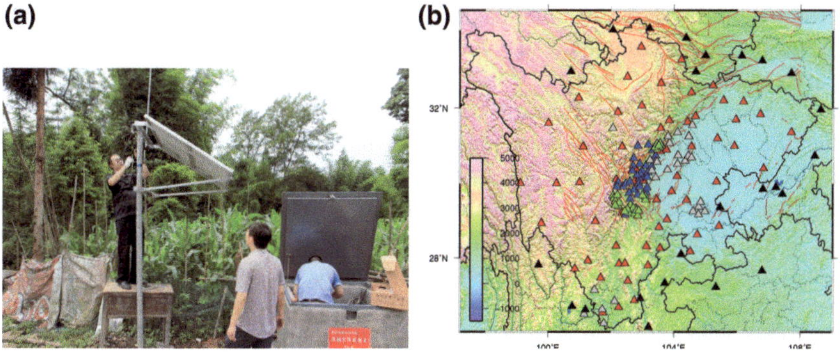

Fig. 2.5 **a** Installation of the temporal seismic stations. From the picture the vault for seismic sensors (*right*) and the solar power panel (*left*) can be seen. Picture by the Portable seismic array group of the field investigation of the 2013 Lushan earthquake. **b** Joint observation and monitoring by the 35 temporal seismic stations (in *blue*), the 60 local and regional seismic stations of Sichuan Province (in *red*), the 32 reservoir seismic stations (in *green*), the 33 public/industry seismic stations (in *gray*), and the 40 seismic stations sharing data with neighboring provinces (in *black*). The data was transmitted in real time to the data center of the earthquake administration of Sichuan Province. Courtesy of Su JR (Earthquake Administration of Sichuan Province)

monitoring. Partly due to this collaborative nature, it might be interesting to notice that in the publications by the team members of the Field Investigation, only a few were acknowledged that the work was supported by the Field Investigation. This, together with the fast and numerous publications mentioned in Chap. 1 and Appendix A1 of this Brief, reflects one side of the characteristics of earthquake science and technology in China, that funding condition seemed good for scientific researches, albeit how to make full use of this unprecedented opportunity to enhance the quality of scientific researches so that such funding state may be sustainable remains a question to be seriously considered by the Chinese institutions.

In the emergency stage, partly due to the lessons and experiences of the Wenchuan earthquake, many scientists and engineers from different agencies and institutions rushed to the disastrous region. However, because the Wenchuan earthquake, with magnitude 8.0 according to the Chinese surface wave magnitude, belongs to the case of 'up-scaling' which showed the insufficient task forces comparing to the natural disasters, the lessons and experiences of this earthquake was not valid to the Lushan earthquake. As a matter of fact, the second day after the earthquake, the CEA, as well as other agencies responsible for the emergency field work, had to announce that without special ratification going to the field was no more permitted. Reason? Transportation blogged by landslides, rock bursts, and, more importantly, traffic jam because of the unbalance between the carry-on capability of the road and the transportation load. This lesson taught people that meeting the challenge of an earthquake, it is better to have a well designed, organized, and well prepared plan for the field work, even if it is not directly the rescue and relief action.

In the emergency stage, the scientific field investigation was mainly coordinated by the headquarter commanding the rescue and relief actions. With the emergency stage completed, the Field Investigation established its own headquarter coordinating the field works and the communication with local authorities. Transportation was shown to be the first difficulty to be overcome, because after the earthquake there were many rock bursts and/or landslides destroying or threatening the transportation. In some regions the local authority had special emergency traffic control for the safety of the travelers along the roads. Rainfall and flood were another threatening factor to be cautioned during the field investigation. Dealing with this special condition, the headquarter of the Field Investigation established hotline communication with the local weather agencies. The headquarter also kept good communication with local government, traffic administrative agencies, and the public. For the quality control purposes, some of the field works were video framed with interview with the investigators, which serves both as a technical archive and as the raw materials for further public outreach.

Time-dependent seismic hazard of the southern Longmenshan fault zone kept to be the main scientific concern of the Field Investigation. Coping with this complex problem, the Field Investigation organized several academic exchanges discussing on the seismic hazard in the Lushan region. In the 8th International Workshop on Statistical Seismology (Statsei8, Beijing), a special session on the Lushan earthquake, together with a panel discussion, was organized. The on-going progresses of the Field Investigation were also introduced at the East-Asia Earthquake Seminar 2013 (Changbaishan, Jilin, by keynote presentation), the Joint China-India Workshop on Earthquake Disaster Mitigation (Shanghai, by special session), and the Cross-Taiwan-Strait Workshop on Earthquake Prediction Experiment Site (Dali, Yunnan, by invited presentation), as well as the Annual Meeting of the Chinese Geophysical Society (Kunming, Yunnan, by co-organized session).

2.3 Concluding Remarks

To understand the scientific problems associated with the Lushan earthquake, namely the seismogenic fault of the Lushan earthquake, the relation between the Lushan earthquake and the Wenchuan earthquake, and the time-dependent seismic hazard of the southern Longmenshan fault zone as well as its surrounding regions, a Field Investigation was organized by the CEA. This Brief is just organized by the Team for the Field Investigation, based on part of the results of the Field Investigation. Preliminary results of the Field Investigation deal with the relocation and depth of aftershocks (Fang et al. 2013; Xu et al. 2013a; Sun et al. 2014; Zhang et al. 2014), discussion on the relation between the Lushan earthquake and the Wenchuan earthquake based on the ETAS model (Jia et al. 2014) and seismogenic structure (Qi et al. 2013; Wang et al. 2013; Wen et al. 2013a; Xiong et al. 2013; Xu et al. 2013b; Ye et al. 2013a, 2013b; Zhang et al. 2013; Chen et al. 2014), gravity anomaly in the seismic source region (Shi et al. 2013), engineering damage of the earthquake (Dai et al. 2013; Sun et al. 2013), earthquake induced landslides (Xu 2013), and seismic hazard assessments (Liu et al. 2014). For the Field Investigation itself, not only scientific results but also the lessons and experiences of the Field Investigation, its design, planning, and organization, are all useful materials for further study.

References

Chen L, Wang H, Ran Y, Lei S, Li X, Wu F, Ma X, Liu C, Han F (2014) The 2013 Lushan M_S7.0 earthquake: varied seismogenic structure from the 2008 Wenchuan earthquake. Seismol Res Lett 85:34–39. doi:10.1785/0220130109

Dai JW, Qu Z, Zhang CX, Weng XR (2013) Preliminary investigation of seismic damage to two steel space structure during the 2013 Lushan earthquake. Earthq Eng Eng Vib 12(3):497–500. doi:10.1007/s11803-013-0189-6

Fang LH, Wu JP, Wang WL, Lü ZY, Wang CZ, Yang T, Cai Y (2013) Relocation of mainshock and aftershock sequences of M_S7.0 Sichuan Lushan earthquake. Chin Sci Bull 58:3451–3459. doi:10.1007/s11434-013-6000-2

Jia K, Zhou SY, Zhuang JC, Jiang CS (2014) Possibility of the independence between the 2013 Lushan earthquake and the 2008 Wenchuan earthquake on Longmenshan fault, Sichuan. Seismol Res Lett 85:60–67. doi:10.1785/0220130115

Liu M, Luo G, Wang H (2014a) The 2013 Lushan earthquake in China tests hazard assessments. Seismol Res Lett 85:40–43. doi:10.1785/0220130117

Liu MJ, Wang FY, Jia SX, Han YJ, Li XM, Feng JL, Sun H, Shen HL, Zhang JS, Tao H, Dai AP, Sun Y, Wang C (2014b) Jianganlin surface fracture induced by 2013 Lushan M_S7.0 earthquake. Acta Seismol Sin 36:129–138. doi:10.3969/j.issn.0253-3782.2014.01.011 (in Chinese with English abstract)

Qi WH, Chen LW, Wen Z, Shan ZD, Wang YL (2013) Preliminary discussion on ground and foundation damages in Ms7.0 Lushan earthquake. J Nat Disasters 22(5):98–103 (in Chinese with English abstract)

Shi L, Chen S, Jiang CS, Xu WM, Lu HY, Guo FY (2013) Characteristics of gravity anomaly in Lushan earthquake zone based on preferential filtering method. Acta Seismol Sin 35:704–716. doi:10.3969/j.issn.0253-3782.2013.05.009 (in Chinese with English abstract)

Sun BT, Wang MZ, Yan PL, Zhang HY, Chen XZ (2013) Damage characteristics and seismic analysis of single-story brick bent frame column industrial buildings in Lushan M_S7.0 earthquake. J Earthq Eng Eng Vib 33(3):1–8. doi:10.11810/1000-1301.20130301 (in Chinese with English abstract)

Sun Z, Wu JP, Fang LH, Wang WL, Wang CZ, Yang T (2014) Focal depth determination of aftershocks of Lushan Ms7.0 earthquake from sPn phase. Chin J Geophys 57(2):430–440. doi: 10.6038/cjg20140209 (in Chinese with English abstract)

Wang YM, Xing LH, Xu WX (2013) Seismic damage and damage enlightenment of medical buildings in Lushan M_S7.0 earthquake. J Earthq Eng Eng Vib 33(4):44–53. doi:10.11810/1000-1301.20130406 (in Chinese with English abstract)

Wen RZ, Ren YF, Huang XT, Lu T, Qi WH (2013a) Strong motion records and their engineering damage implications for Lushan earthquake on April 20, 2013. J Earthq Eng Eng Vib 33(4):1–14. doi: 10.11810/1000-1301.20130401 (in Chinese with English abstract)

Xiong LH, Lan RQ, Wang YM, Tian XM, Feng B (2013) Earthquake damage investigation of structures in 7.0 Lushan strong earthquake. J Earthq Eng Eng Vib 33(4):35–43. doi:10.11810/1000-1301.20130405 (in Chinese with English abstract)

Xu C (2013) Quick landslide susceptibly evaluation and its validation for the 2013 M_S7.0 Lushan earthquake. Sci Technol Rev 31(28/29): 15–23. doi: 10.3981/j.issn.1000-7857.2013.h2.001 (in Chinese with English abstract)

Xu LS, Yan C, Zhang X, Li CL (2013a) Where did the Lushan M_S7.0 earthquake occur in the world? Chin J Geophys 56:2982–2993. doi:10.6038/cjg20130912 (in Chinese with English abstract)

Xu XW, Wen XZ, Ye JQ, Ma BQ, Chen J, Zhou RJ, He HL, Tian QJ, He YL, Wand ZC, Sun ZM, Feng XJ, Yu GH, Chen LC, Chen GH, Yu SE, Ran YK, Li XG, Li CX, An YF (2008) The M_s8.0 Wenchuan earthquake surface ruptures and its seismogenic structure. Seismol Geol 30:597–629 (in Chinese with English abstract)

Xu XW, Chen GH, Yu GH, Cheng J, Tan XB, Zhu AL, Wen XZ (2013b) Seismogenic structure of Lushan earthquake and its relationship with Wenchuan earthquake. Earth Sci Front 20(3):11–20 (in Chinese with English abstract)

Ye F, Guo ED, Liu JL, Liu RS, Lin JQ, Gao L (2013a) Seismic damage investigation and analysis of water supply system in Lushan earthquake. World Earthq Eng 29(4):52–58 (in Chinese with English abstract)

Ye F, Guo ED, Liu JL, Liu RS, Lin JQ, Liu Z (2013b) Seismic damage investigation and analysis of gas system in Lushan M_S7.0 earthquake. J Nat Disasters 22(5):77–82 (in Chinese with English abstract)

Zhang LX, Liu C, Liu JP (2013) Investigation and consideration of the staircase damage in Lushan earthquake. J Earthq Eng Eng Vib 33(4):21–28. doi:10.11810/1000-1301.20130403 (in Chinese with English abstract)

Zhang Y, Chen YT, Xu L, Du F, Jin M, Tu H, Dahm T (2014) Kinematic rupture model and hypocenter relocation of the 2013 M_W6.6 Lushan earthquake constrained by strong-motion and teleseismic data. Seismol Res Lett 85:15–22. doi: 10.1785/0220130126

Chapter 3
Tectonic Setting of the Earthquake

Abstract The seismo-tectonic picture of the Lushan earthquake is clear at the scales of lithospheric plate and crustal block, but remains puzzle at the scale of seismogenic faults. Role of such 'missing earthquakes' in seismic hazard analysis is a question worth discussing.

Keywords Tectonic setting · Seismogenic fault · The Longmenshan fault zone · Tibetan plateau

3.1 Geodynamic Background

The occurrence of the Wenchuan earthquake and the Lushan earthquake is closely related to the interaction between the Indian plate and the Eurasian plate which forms the Tibetan plateau (Fig. 3.1, see Burchfiel et al. 2008). At a larger spatial scale, major to great earthquakes in China are tectonically determined by the interaction among several surrounding tectonic plates, including the Euroasia, the India, the Philippines, the North America, and the Pacific plate. At a smaller spatial scale, the crust of continental China can be divided into several 'tectonic blocks', such as the South China block, the Bayan Har block, and so on. Most of the major, and all of the great earthquakes on record distributed along the block boundary zones (Zhang et al. 2003).

As one of the 'active' tectonic blocks, the Bayan Har block is located in the central and eastern part of the Tibetan plateau. The Longmenshan thrust belt is the eastern margin of the Bayan Har block. From 1997 to the time when this manuscript was finished, the Bayan Har block has been in a seismically active period. A series of major to great earthquakes have occurred along the boundary zone of this tectonic block, being the unique contributor to the major to great earthquakes in the Chinese mainland in this period (see Figs. 3.2 and 3.3). Although the tectonic picture can clearly explain

---

This chapter is written by Ding ZF, Wang XC and Jiang CS (Institute of Geophysics, CEA).

Z. Wu et al., *Earthquake Phenomenology from the Field*, SpringerBriefs in Earth Sciences, DOI: 10.1007/978-981-4585-15-6_3,

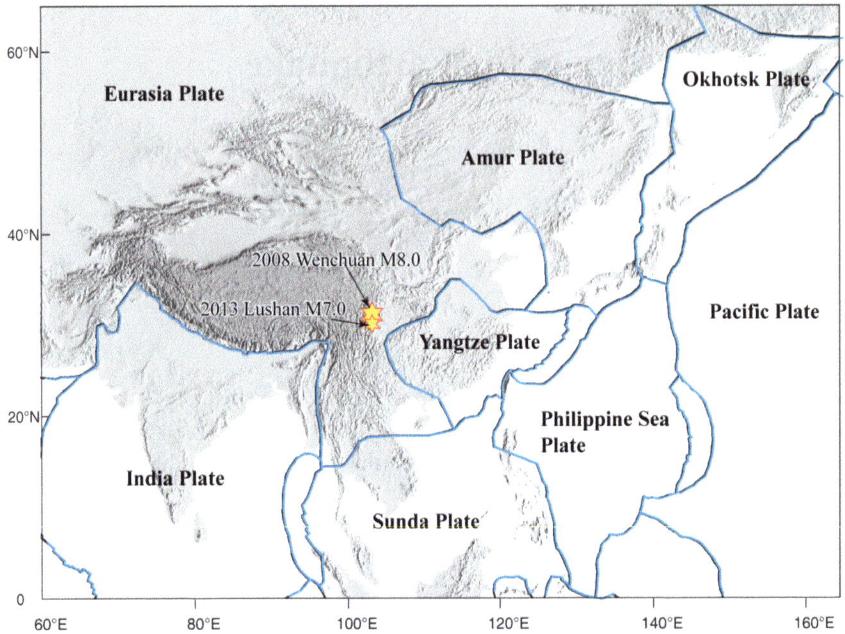

Fig. 3.1 The 2008 Wenchuan M_S8.0 earthquake and the 2013 Lushan M_S7.0 earthquake, in the context of the plate tectonics (Bird 2003) around China. From the topography the Tibetan plateau can be seen, in between the India plate and the Eurasia plate

the focal mechanisms of these major to great earthquakes, the geodynamic reason why there is such a temporal clustering behavior remains unknown.

The convergence rate across the Longmenshan mountain is low, and great earthquakes are rare (Gan et al. 2007). Paleoseismic works that have been done since the Wenchuan earthquake revealed several past great earthquakes of similar size to the Wenchuan earthquake in 2008, with the recurrence period 2000–3000 years. Assessing the seismic hazard along such a fault zone, therefore, one has to consider both moment release and moment accumulation at the same time, that is, the moment deficit (Wang et al. 2010).

3.2 The Seismo-Tectonics of the Lushan Earthquake: Different Views, and Their Evidences

3.2.1 Surface Rupture and Active Faults

After the Lushan earthquake, Field Investigation was carried out aiming at the seismo-tectonic model of the earthquake. Before the earthquake, supported by the local government of Ya'an City, geologists from the Institute of Geology, China Earthquake Administration (CEA) carried out an active fault mapping project. The

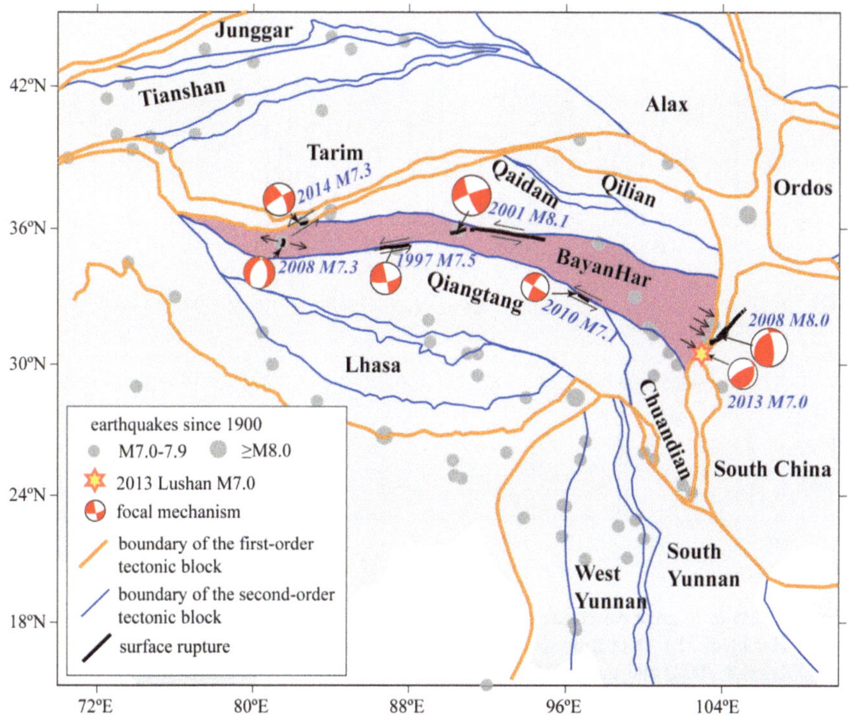

Fig. 3.2 The tectonic blocks (with names of the blocks shown by the text) in west China and its surrounding regions, with the Bayan Har block highlighted, along with the major to great earthquakes around its boundary since 1997. The tectonic block boundaries are from Zhang et al. (2003). Parameters of the earthquakes (*gray dots*, with different sizes indicating the magnitude range) are from the IASPEI global catalogue 1900–1999 (Engdahl and Villaseñor 2002) and the GCMT catalogue since 2000. To the west of the Wenchuan earthquake and the Lushan earthquake is the west-Sichuan plateau (belonging to the Bayan Har block); To the east of the two earthquakes is the Chengdu Basin (belonging to the south China block); The Longmenshan fault zone spans in between, accommodating the Wenchuan earthquake and the Lushan earthquake

project was just going on when the earthquake occurred (Chen et al. 2013a). This is, to much extent, an ironic lesson for earthquake science. In the perspective of science, people generally have some 'feelings' of the potentially dangerous regions in need of enhanced observation, monitoring and survey. Problem is that the resolution of modern earthquake science as per the time of the 'target' earthquake is low, and the earthquake sometimes runs ahead of the scientific work. A similar case was just nearby and just a few years ago: From 2005, the Ministry of Science and Technology (MOST) of China supported a national project (2005–2010) on the geodynamics and earthquakes in the western Sichuan region. The project deployed a seismic array with 200 mobile broadband seismic stations, and captured in near source distance the mainshock (and aftershocks) of the Wenchuan earthquake which was not forecasted before hand.

In the Field Investigation of the Lushan earthquake, geological works are mainly focused both on the active fault mapping and on the finding of the surface rupture caused by the earthquake. Although the survey of different agencies

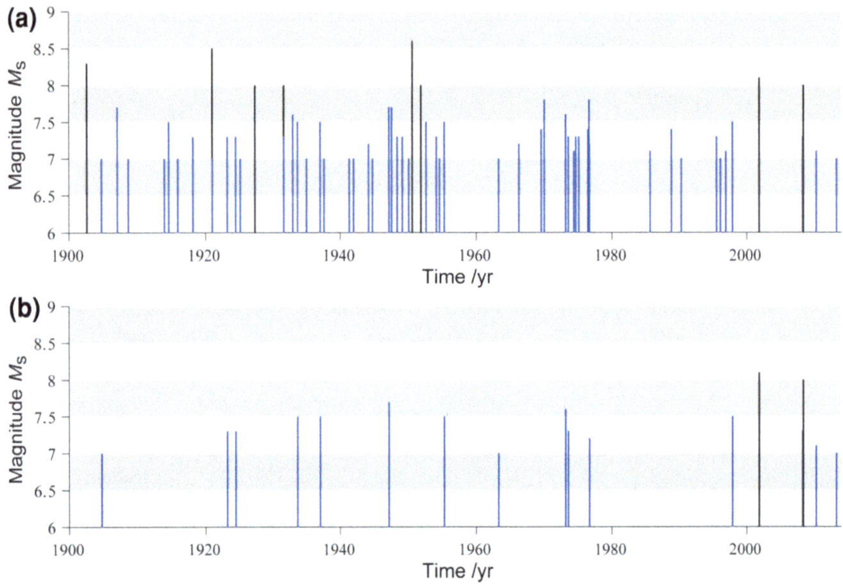

Fig. 3.3 Major to great earthquakes in the Chinese mainland (**a**) and along the boundary of the Bayan Har block (**b**). Data from the IASPEI global earthquake catalogue 1900–1999 (Engdahl and Villaseñor 2002) and the GCMT catalogue since 2000. It may be seen that since 1997 the Bayan Har block has been acting as the unique contributor of major to great earthquakes to the Chinese mainland

found no primary surface faulting caused by the Lushan earthquake, the active fault mapping and paleo-seismological studies in Ya'an City still provided the understanding of the southern Longmenshan fault zone with useful clues. By field investigation and comparison with the Wenchuan earthquake, Chen et al. (2013a) concluded that, different from the Wenchuan earthquake which occurred mainly in the central to front-range fault system, the Lushan earthquake occurred in the front-range to range-front fault system, which is to the east of the former. Combing field survey and GPS measurement Li et al. (2013a) obtained similar conclusion that the seismogenic structure of the Lushan earthquake lies between the Shuangshi-Dachuan fault (one segment, or the southern extension, of the front-range fault) and the Xinkaidian fault (one segment, or the southern extension, of the range-front fault).

3.2.2 Deep Structure of the Vicinity of the Seismic Source

The tectonic context that the western Sichuan plateau is being thrust over the Chengdu Basin at the Longmenshan fault zone (Burchfiel et al. 2008) determined the seismogenesis of the Wenchuan earthquake, and the Lushan earthquake as well. Field investigation and geophysical studies of the Wenchuan earthquake described the earthquake preparation process by a ‘three units model’ (Chen and Booth 2011): the deformation unit (the western Sichuan plateau), the locking unit (the Longmenshan fault zone), and the resisting unit (the Chengdu Basin), which seems also valid for the Lushan earthquake, albeit not identical in details.

Therefore, seen along the east-west profile, the problem of the structure of the source region of the Lushan earthquake is similar to that of the Wenchuan earthquake: the clear side is the seemingly abrupt change of crust thickness and the ‘three units’. Gravity and topography give that the mechanical strength of the west-Sichuan plateau and the Chengdu Basin is different (Chen et al. 2013b). According to Shen (2013) using teleseismic receiver functions, even the lithosphere-asthenosphere boundary (LAB) has abrupt changes crossing the source region of the Lushan earthquake. Moreover, MT data shows that the Longmenshan fault zone is a high-resistivity body (HRB) extending from the ground surface to the upper mantle, which might be the indication of the ‘locking unit’ or ‘rupturing unit’ (Zhan et al. 2013). Meanwhile, the unclear side is the structure and property of both the deep fault and (the existence of) the brittle-ductile transition zone in the crust as well as the detachment layer, which remains a scientific target of higher-resolution geophysical mapping. On the other hand, however, one of the key issues to be considered at this moment is the segmentation of the southern Longmenshan fault zone which may determine the future seismic hazard in that region. Remarkably, how to understand the two ‘seismic gaps’ on each side of the Lushan aftershock sequence is directly related to this scientific problem.

In studying the structure of the seismic source, geophysical measures always have the problem of limited resolution. Therefore, like detective collecting evidences, it is important to get useful clues from the evidences with large uncertainties and with limited detection capability. At present time, the study is still going on. But the preliminary results of Li et al. (2013c) using local and near source seismic stations recording the aftershocks, and Zheng et al. (2013) by applying the NCF method and the teleseismic receiver function method to the national and regional seismic networks both indicate a clear segmentation picture that the whole southern Longmenshan fault zone can be divided into three parts by the physical properties of the Earth medium such as Poisson ratio or velocity anomaly, with the 2013 Lushan earthquake occurred in the middle segment. This gives a clear indication of the maximum magnitude of future earthquakes in a ‘normal situation’, in which the stressing of the ‘normal situation’ is to caution the case that different segments be connected to each other to form a Sumatra-Andaman-like ‘extreme event’.

It may not be making sense in science to have a look at the names of places in that region. Along the mid to north Longmenshan range there are Wenchuan, Beichuan, and Qingchuan, in which the Chinese character Chuan = river. To the south, in contrast, there are Lushan, Mingshan, E’meishan, and Leshan, in which the Chinese character Shan = mountain. This can be regarded as a ‘culture sign’ of the segmentation of the Longmenshan fault zone.

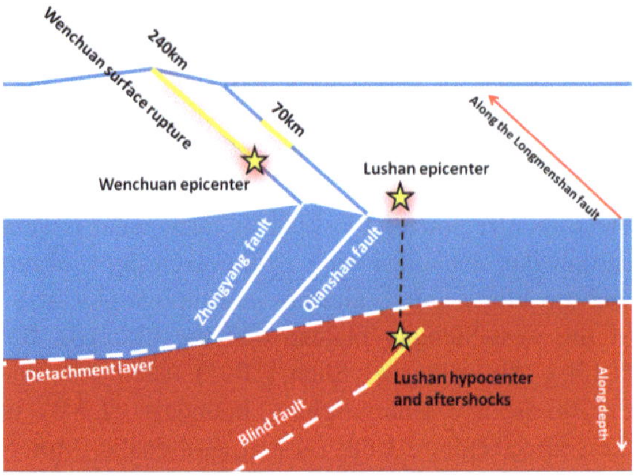

Fig. 3.4 Sketch of the 'blind fault version' of the tectonic model of the 2013 Lushan earthquake, with the reference of the 2008 Wenchuan earthquake. The data is based on Chen and Booth (2011) and Xu et al. (2013b)

3.2.3 The Lushan Puzzle

Almost all the geological groups concluded that there is no primary surface rupture found associated with the Lushan earthquake. What is the real cause of the Lushan earthquake, therefore, remains puzzle and controversial, partly due to this situation. Li et al. (2013b) and Zhang et al. (2013) associated the Lushan earthquake with the Shuangshi-Dachuan fault (the extension of the front-range fault). Based on geological survey, Han et al. (2013) discussed the 'surface rupture signs' and pointed out that, although there is no surface rupture on the ground, the distribution of such 'surface rupture signs' shows that a hidden fault (the 'Lushan-Longmen presumed blind fault') may be responsible for the Lushan earthquake. Xu et al. (2013a, b) used geological and geodetic evidences to demonstrate that the Lushan earthquake may be caused by a blind reverse fault. The upper tip of this blind fault is below the depth 9 km. All the above hypotheses were based on the field works in the early stage of the Field Investigation. Figure 3.4 shows the sketch of the 'blind fault version' of the tectonic model of the Lushan earthquake and the relation with the 2008 Wenchuan earthquake.

3.3 Summary and Discussion

3.3.1 The Blind-Lame Combination of Seismology and Tectonics

Associate of a seismogenic fault with an earthquake is a typical kind of blind-lame cooperation between seismology and tectonics. Figure 3.5 shows two results by different groups about the same region, the same fault, and the same earthquake (Xu et al. 2013a, b; Zhang et al. 2013). From each figure the relation between the earthquake and the fault is clear, albeit both results have to consider the dual

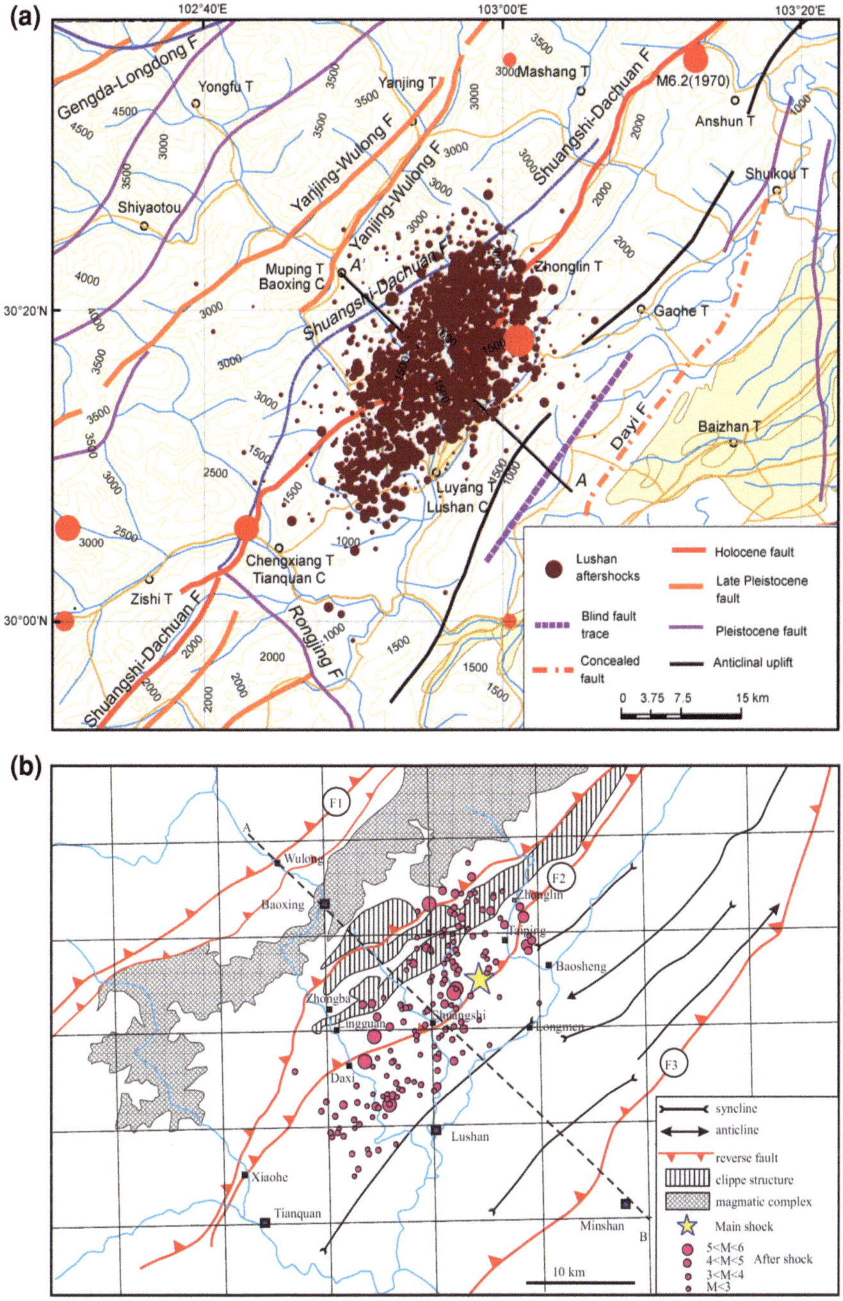

Fig. 3.5 The Lushan earthquake and the Shuangshi-Dachuan fault, maps from different groups: **a** Xu et al. (2013b); **b** Zhang et al. (2013)

uncertainty of the comparison: one comes from the position of the fault for which geologists are responsible, while the other comes from the location of the earthquake for which seismologists are responsible. Fortunately for the Lushan earthquake, the government of Ya'an City sponsored the active fault mapping project which helped in constraining the exact position of the Shuangshi-Dachuan fault, and the local seismic stations constrained the locations of the aftershocks, and in turn, the location of the mainshock by their relative locations. By the updated results of active fault mapping and the relocation of the earthquakes, it seems that the 'blind fault version' of the seismo-tectonic model is more convincing. Probably it is still too early to reach the definite conclusion, since the work is still going on.

3.3.2 The Role of the 'Missing Earthquakes' in Seismic Hazard Assessment

Suppose that we were in the ancient time. We did not have the written documentations of the Lushan earthquake, because in Lushan and its vicinity there was almost no population. Neither did we have seismological recordings because at that time seismographs had not been invented. Then because there was no surface rupture of this M_S7.0 earthquake, our descendant geologists would never have the information of this earthquake. That is, this earthquake, which did have occurred in the history along the Longmenshan fault, would become a 'missing event' in our scientific/historical recordings. Accordingly, in the assessment of seismic hazard of our descendants, if they use the similar knowledge like what we are using today, there would be problems caused by the 'missing' of such kind of earthquakes.

This scenario is by no means a science fiction. It is well-known that many strong to major earthquakes have no primary surface ruptures. Another example is the 2001 Bhuj earthquake. In the perspective of statistical seismology, how to evaluate the effect of such 'missing earthquakes' on the assessment of seismic hazard is one of the questions not only relevant in science but also of practical significance to social sustainability—although there has not been a definite answer yet.

References

Bird P (2003) An updated digital model of plate boundaries. Geochem Geophys Geosyst 4(3):1027. doi:10.1029/2001GC000252

Burchfiel BC, Royden LH, Vander RD, Hager BH, Chen Z, King RW, Li C, Hao LJ, Kirby E (2008) A geological and geophysical context for the Wenchuan earthquake of 12 May 2008, Sichuan. People's Republic of China. GSA Today 18(7):4–11. doi:10.1130/GSATG18A.1

Chen LC, Ran YK, Wang H, Li YB, Ma XQ (2013a) The Lushan M_S7.0 earthquake and activity of the southern segment of the Longmenshan fault zone. Chin Sci Bull 58:3475–3482. doi:10.1007/s11434-013-6009-6

Chen S, Xu WM, Shi L, Lu HY, Guo FY (2013b) Gravity field and lithospheric mechanical properties of Longmenshan fault zone and its surrounding areas. Acta Seismol Sin 35:692–703. doi:10.3969/j.issn.0253-3728.2013.05.008 (in Chinese with English abstract)

Chen Y, Booth DC (2011) The Wenchuan earthquake of 2008: anatomy of a disaster. Science Press in cooperation with Springer, Beijing

Engdahl ER, Villaseñor A (2002) Global seismicity: 1900–1999. In: Lee WHK, Kanamori H, Jennings PC, Kisslinger C (eds) International handbook of earthquake and engineering seismology, Part A. Academic Press, San Diego, pp 665–690

Gan WJ, Zhang PZ, Shen ZK, Niu ZJ, Wang M, Wan YG, Zhou DM, Cheng J (2007) Present-day crustal motion within the Tibetan plateau inferred from GPS measurements. J Geophys Res 112:B08416. doi:10.1029/2005JB004120

Han ZJ, Ren ZK, Wang H, Wang MM (2013) The surface rupture signs of the Lushan "4·20" M_S7.0 earthquake at Longmen township, Lushan county and its discussion. Seismolog Geol 35:388–397. doi:10.3969/j.issn.0253-4967.2013.02.017 (in Chinese with English abstract)

Li CY, Xu XW, Gan WJ, Wen XZ, Zheng WJ, Wei ZY, Xu C, Tan XB, Chen GH, Liang MJ, Li XN (2013a) Seismogenic structures associated with the 20 April 2013 M_S7.0 Lushan earthquake Sichuan Province. Seismolog Geol 35:671–683. doi:10.3969/j.issn.0253-4967.2013.03.020 (in Chinese with English abstract)

Li YS, Wang YS, Pei XJ, Luo YH (2013b) Research on tectonic fracturing and causative fault of Lushan earthquake in Sichuan, China. J Chengdu Univ Technol (Sci Technol edn) 40:242–249. doi:10.3969/j.issn.1671-9727.2013.03.03 (in Chinese with English abstract)

Li ZW, Tian BF, Liu S, Yang JS (2013c) Asperity of the 2013 Lushan earthquake in the eastern margin of Tibetan plateau from seismic tomography and aftershock relocation. Geophys J Int 195:2016–2022. doi:10.1093/gji/ggt370

Shen XZ (2013) An analysis of the deformation of the crust and LAB beneath the Lushan and Wenchuan earthquakes in Sichuan Province. Chin J Geophys 56:1895–1903. doi:10.6038/cjg20130612 (in Chinese with English abstract)

Wang H, Liu M, Shen XH, Liu J (2010) Balance of seismic moment in the Songpan-Ganze region, eastern Tibet: Implication for the 2008 great Wenchuan earthquake. Tectonophysics 491:154–164. doi:10.1016/j.tecto.2009.09.022

Xu XW, Chen GH, Yu GH, Cheng J, Tan XB, Zhu AL, Wen XZ (2013a) Seismogenic structure of Lushan earthquake and its relationship with Wenchuan earthquake. Earth Sci Front 20(3):11–20 (in Chinese with English abstract)

Xu XW, Wen XZ, Han ZJ, Chen GH, Li CY, Zheng WJ, Zhang SM, Ren ZK, Xu C, Tan XB, Wei ZY, Wang MM, Ren JJ, He Z, Liang MJ (2013b) Lushan M_S7.0 earthquake: a blind reserve-fault earthquake. Chin Sci Bull 58:3437–3443. doi:10.1007/s11434-013-5999-4

Zhan Y, Zhao GZ, Unsworth M, Wang LF, Chen XB, Li T, Xiao QB, Wang JJ, Tang J, Cai JT, Wang YZ (2013) Deep structure beneath the southwestern section of the Longmenshan fault zone and seismogenetic context of the 4.20 Lushan M_S7.0 earthquake. Chin Sci Bull 58:3467–3474. doi:10.1007/s11434-013-6013-x

Zhang PZ, Deng QD, Zhang GM, Ma J, Gan VJ, Min W, Mao FY, Wang Q (2003) Active tectonic blocks and strong earthquakes in the continent of China. Sci Chin Ser D: Earth Sci 46:13–24. doi:10.1360/03dz0002

Zhang YQ, Dong SW, Hou CT, Shi JS, Wu ZH, Li HL, Sun P, Liu J (2013) Preliminary study on the seismotectonies of the 2013 Lushan M_S7.0 earthquake west Sichuan. Acta Geol Sinica 87:447–458

Zheng Y, Ge C, Xie ZJ, Yang YJ, Xiong X, Xu HZ (2013) Crustal and upper mantle structure and the deep seismogenic environment in the source regions of the Lushan earthquake and the Wenchuan earthquake. Sci Chin: Earth Sci 56:1158–1168. doi:10.1007/s11430-013-4641-2

Chapter 4
Aftershock Sequence: Monitoring and Analysis

Abstract Based on the regional and local seismic networks, especially the temporary mobile seismic stations deployed in the epicentral region, the aftershock sequence of the Lushan earthquake was well constrained. ETAS model was applied to the analysis of the aftershock activity, approaching to the seismological information service facilitating the rescue and relief actions as well as the scientific investigation of the earthquake.

Keywords Aftershock sequence · Double difference (DD) relocation · The Epidemic-type Aftershock Sequence (ETAS) model

4.1 Introduction

Aftershock sequence of an earthquake can provide important information about the seismic source process of the earthquake and the structure of the source region. Monitoring of aftershock activity has direct importance for the rescue and relief actions, the reconstruction, and the field work of the scientific investigation. As a part of the Field Investigation of the 2013 Lushan earthquake, monitoring of the aftershock sequence, the relocation (Fang et al. 2013; Zhang and Lei 2013), focal mechanism determination (Lü et al. 2013; Zhao et al. 2013a; Han et al. 2014), and the analysis of the sequence parameters based on the ETAS model (Jiang et al. 2013) provided rich data for further studies. Aftershock recordings were also used to retrieve the structure of the source region (Li et al. 2013). These results, together with the focal mechanism (Zeng et al. 2013a, b) and rupture process (Liu et al. 2013; Wang et al. 2013; Xie et al. 2013; Zhang et al. 2013; Zhao et al. 2013b) of the mainshock, and the relation with the stress change caused by the mainshock (Miao and Zhu 2013), provided evidences for the understanding of the source process of the mainshock and the generation of

---

This chapter is written by Fang LH and Jiang CS (Institute of Geophysics, CEA), with the assistance of Zhang SF and Jiang H.

Z. Wu et al., *Earthquake Phenomenology from the Field*, SpringerBriefs in Earth Sciences, DOI: 10.1007/978-981-4585-15-6_4, © The Author(s) 2014

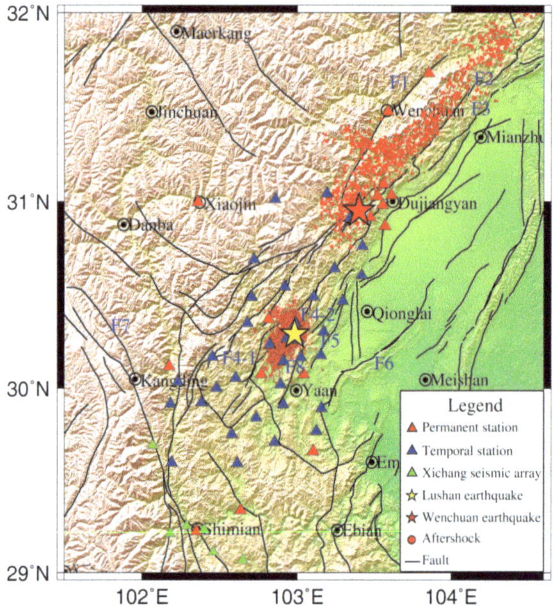

Fig. 4.1 Distribution of seismic stations near the epicenter, with faults and seismicity as background. The main active faults are: *F1* Wenchuan-Maowen fault, *F2* Beichuan-Yingxiu fault, *F3* Jiangyou-Guanxian fault, *F4-1* Shuangshi-Dachuan main fault, *F4-2* Shuangshi-Dachuan branch fault, *F5* Dayi fault, *F6* Pujiang-Xinjin fault, *F7* Xianshuihe fault, *Fx* unknown fault

the aftershocks, and provided heuristic clues to the seismic hazard of the 'seismic gap' between the Lushan earthquake sequence and the Wenchuan earthquake sequence (Gao et al. 2013).

4.2 Aftershock Monitoring

In an effort to monitor the seismic activity of the 'seismic gap' between the Lushan earthquake sequence and the Wenchuan earthquake sequence and enhance the observations of the Lushan aftershock sequence, the Institute of Geophysics, China Earthquake Administration (CEA) deployed 35 temporary mobile seismic stations around the epicentral area. Each seismic station was comprised of a 24-bit recorder (REFTEK/EDAS-24IP) with a GPS timing system. Thirteen of these stations were equipped with short-period seismometers (CMG-40T) and 22 were equipped with broadband sensors (CMG-3ESPC and CMG-3T). Data was recorded at 100 sps in a continuous mode. Besides, waveform data recorded by national and regional seismic networks were collected. The Xichang Seismic Array, which was deployed by the Institute of Geophysics, CEA, to monitor the local seismic activity, also joined in the monitoring. Figure 4.1 shows the distribution of seismic stations used to monitor the aftershocks. The seismic stations cover the aftershock region with an average azimuthal gap less than 150°. The distances between the stations and the epicenters vary from 0.1 to 200 km.

By the end of 24:00:00 (local time) on 3 December, 2013, 13,811 aftershocks were registered by this joint temporary seismic array. The magnitude of

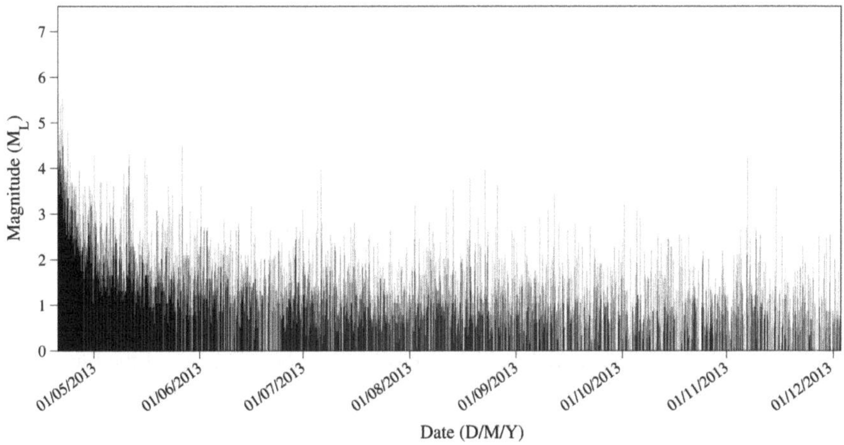

Fig. 4.2 The magnitude-time plot of the aftershock sequence, in the period from April 20 to December 4, 2013

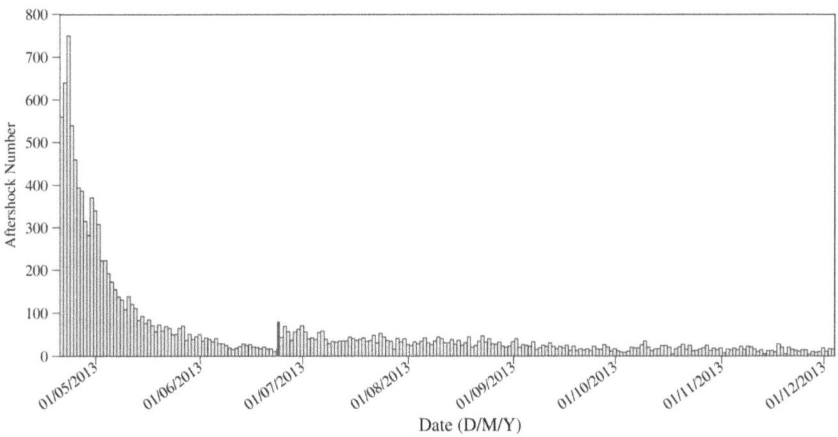

Fig. 4.3 Number of recorded aftershocks per day in the period from April 20 to December 4, 2013, with the *red vertical line* indicating the significant enhancement of the monitoring capability due to the deployment of the mobile seismic stations

the aftershocks ranges from M_L −0.5 to M_L5.6. There are 6 aftershocks with $M_L \geq 5.0$ and 54 aftershocks with $M_L \geq 4.0$. Figure 4.2 shows the magnitude-time diagram of the aftershock sequence. Figure 4.3 shows the enhancement of the monitoring capability by deploying the mobile seismic network and the joint analysis of the aftershocks. Note that when the 2001 Kokoxili, Qinghai, M_S8.1 earthquake occurred, the aftershock monitoring used only a few analogue seismic stations. The advancement of seismic observations was to much extent a result of the on-going capacity building endeavor since the turn of the centuries.

4.3 Relocation of the Aftershocks

Analyzing the aftershock sequence, the waveform data were transmitted to the Earthquake Administration of Sichuan Province in real time, which were subject to further analysis. P and S arrivals were manually picked for all stations, in which P phases were picked on the vertical component seismograms and S waves on the horizontal component seismograms. There are totally 99,413 P arrivals and 98,485 S arrivals picked for the 13,811 events. On average, each event has 15 phase readings.

Initial event locations were obtained using program LOC3D (Wu et al. 2009; Fang et al. 2011, 2013). The program calculates theoretical arrival times of regional phases based on the 3-D velocity model and takes into account the ellipticity of the Earth, station elevation, and topography. Cross-correlation was applied for the accurate reading of the travel time differences of the P- and S- phase arrivals from different earthquakes. The relocation of the aftershocks used the double-difference (DD) algorithm HYPODD (Waldhauser and Ellsworth 2000), incorporating these travel time differences, and minimizing the location errors caused by the un-modeled velocity structure. A total of 12 iterations for the conjugate gradient method (LSQR) within the HYPODD inversion were taken. The travel time differences have been estimated for all the event pairs with an inter-event separation less than 6 km and stations located within 150 km radius from the cluster centroid. A maximum of 8 neighboring events linked to each other were considered for the relocation. The condition numbers (i.e. ratio of the largest to the smallest eigen value) obtained for the 12 iterations range from 45 to 81. The *a priori* weights assigned for the P and S arrivals were 1.0 and 0.5, respectively. The average 'relative uncertainties' (equivalent to the precision) for the aftershocks upon relocation are of the order of 0.05 km in epicentral location and 0.07 km in focal depth estimation. As shown by Fig. 4.4, location errors were significantly reduced by both the double-difference location algorithm and the cross-correlation of seismic waveforms.

Figure 4.5 shows the relocated epicenter distribution. From the figure it may be seen that the aftershocks are distributed mainly in the NNE direction along both sides of the Shuangshi-Dachuan fault. A few earthquakes occurred in the southern part of aftershock zone but more in the middle-northern part. The width of the aftershock swarm changes near Lushan, being wide in the north of Lushan and narrow in the south. Most aftershocks are mainly in the Mesozoic and Cenozoic basins and a few in the mountain area.

The relocation results show that the aftershocks spread approximately 45 km in length and 20 km in width. The dominant distribution of the focal depth ranges from 10 to 20 km. A few earthquakes occurred in the shallow crust. Focal depth profiles show fault planes dip to the northwest. The dip angle of the seismogenic fault is about 41°. Although seen on the surface map, the epicenters of aftershocks distributed mainly along both sides of the Shuangshi-Dachuan fault, seen in a three-dimensional perspective, the seismogenic fault may be a blind thrust fault on the eastern side of the Shuangshi-Dachuan fault. The relocation results also reveal that there is a southeastward tilt aftershock belt intersecting with the seismogenic fault with *Y*-shape. It might be inferred that this is a back thrust fault.

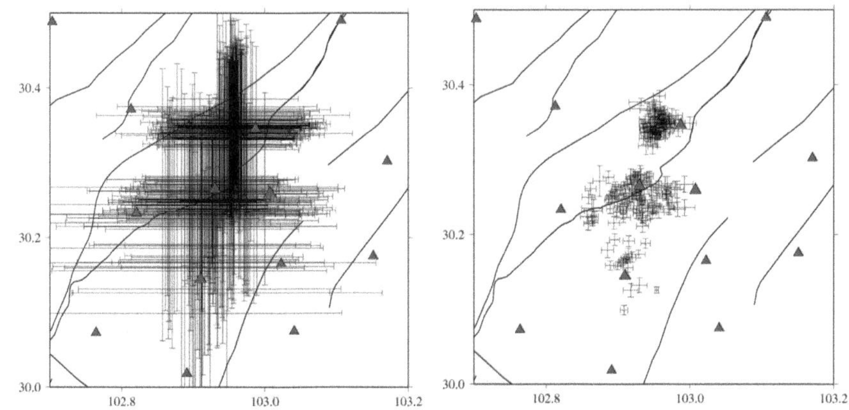

Fig. 4.4 Comparison of location errors with and without waveform cross correlation. *Left* location result using the difference of catalog travel times for each event pair; *Right* location result using the difference of catalog travel times and waveform cross-correlation travel-time differences for each event pair. The *red triangles* represent seismic stations. The seismic events are from 2013/6/26 to 2013/10/5

4.4 Statistical Analysis of Sequence Parameters by Fitting the ETAS Model

4.4.1 The Epidemic-Type Aftershock Sequence Model

Aftershock activity is not always well predicted by the single modified Omori-Utsu formula (Omori 1894; Utsu 1961), especially when it includes the conspicuous secondary aftershock activities of large aftershocks. Ogata (1989, 1992) assumed that every aftershock can trigger further aftershocks or remote events, and extended the modified Omori-Utsu formula to the Epidemic-Type Aftershock Sequence (ETAS) model by using the following conditional intensity function

$$\lambda(t) = \mu + K \sum_{t_i < t} \frac{e^{\alpha(M_{\mathrm{i}} - M_0)}}{(t - t_i + c)^p}, \quad M_i > M_0 \tag{4.1}$$

in which μ is the occurrence rate for the background seismic activity, M_{i} and t_{i} are the magnitude and occurrence time of the ith event, K is a normalizing constant governing the expected number of direct aftershocks triggered by earthquake i, and M_0 is the cutoff magnitude of the catalogue in use.

Comparing to the Omori's law, the p value indicates the decay rate of the aftershock sequence, and the α value measures the efficiency of an earthquake in generating its offspring (Ogata 1989, 1992). These parameters are useful in characterizing local seismicity, in which swarm activities have smaller values. Maximizing the log-likelihood

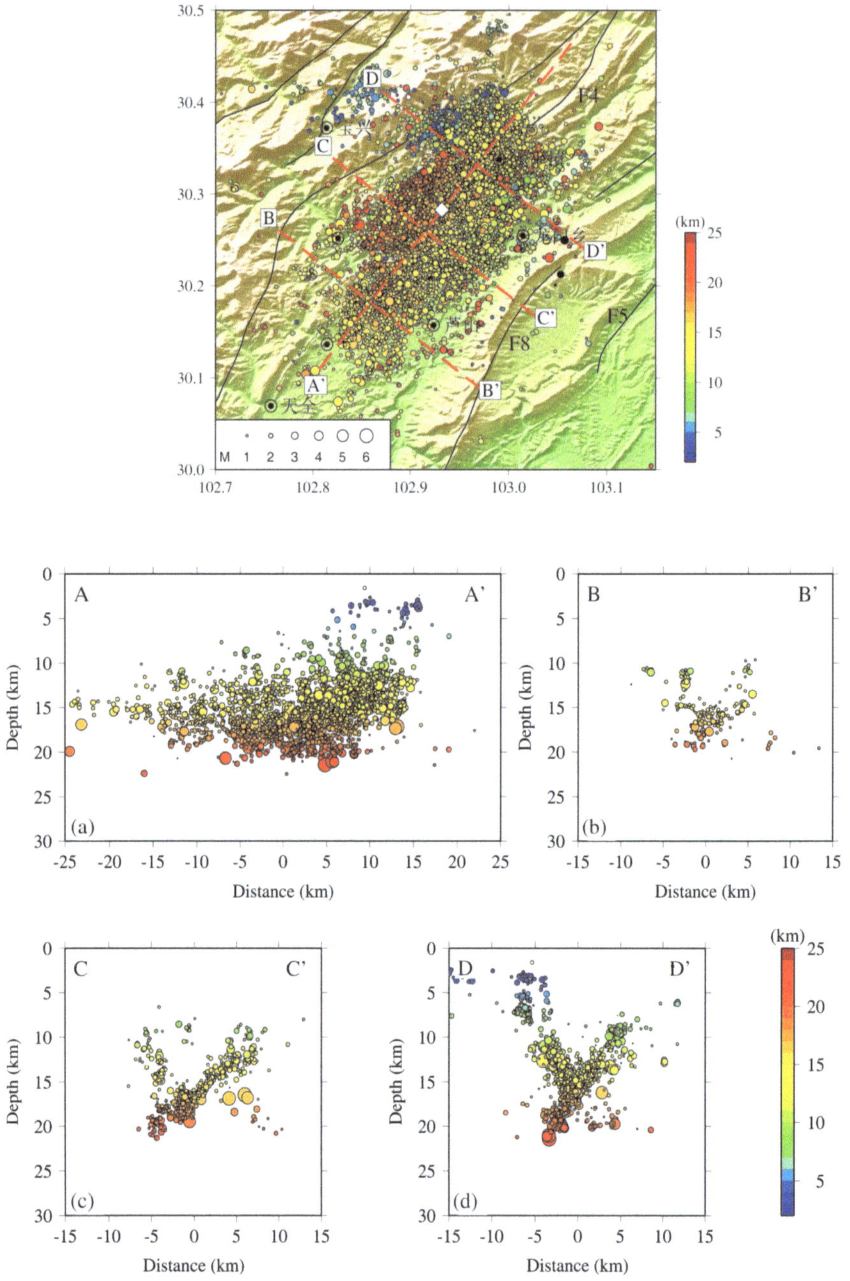

Fig. 4.5 Relocated aftershocks: map view and profile projection, with *colors* indicating the focal depths. The distance between each earthquake and the surface projection of the cross-section is less than 3 km. *Red star* indicates the mainshock. Aftershocks scale with the magnitudes

$$\lg L = \sum_{i:S \le t_i < T} \lg \lambda(t_i) - \int_S^T \lambda(t)dt \tag{4.2}$$

and using the residual point process (RPP) analysis to transform the function $\lambda(t)$ into a stationary Poisson process by the time transposition

$$t \mapsto \tau = \Lambda(t) = \int_0^t \lambda(u)du \tag{4.3}$$

parameters μ, K, c and p can be estimated.

4.4.2 ETAS Parameters in the Early Stage of the Lushan Earthquake Sequence

The Lushan earthquake sequence being still continuing, the analysis of its early stage can provide some information of its overall feature. For this purpose the ETAS model was fitted to the early stage of the aftershock sequence (Jiang et al. 2013), with the cutoff magnitude M_c being M_L2.0 according to the completeness analysis of the earthquake catalogue[1] (see Fig. 4.6). Figure 4.7 shows the temporal variation of the conditional intensity with fitting time interval from 0.31 to 24.12 days, and the parameters obtained by the maximum likelihood estimate. The ETAS model fitting obtains $\alpha = 1.89$, $p = 1.22$, and $b = 0.72$ estimated independently from the G-R relation. Comparing to other earthquake sequences in the continental regions of China (Jiang et al. 2007) and Japan (Guo and Ogata 1997), the parameters indicate that the Lushan earthquake sequence has a relatively weak triggering ability in generating secondary aftershocks, a quick decay rate of aftershocks, and a lower b-value in the aftershock region. Comparing to the result of Wang (1994), the Lushan earthquake falls into the b–p scaling relation, and is closer to the negative regression line comparing to other earthquakes in the Chinese mainland (open squares in Fig. 4.8).

To examine the stability of estimated parameters, different cutoff magnitudes M_c and different ending times of the fitting interval have been considered, as shown in Table 4.1 and Fig. 4.9, which show that M_c affects the value of α significantly, but has less influence on p. Furthermore, the temporal variation of the ETAS parameters have larger fluctuations within the first 10 days after the mainshock in which the number of earthquakes were not sufficient to obtain a stable estimate, and become more stable with time lapses.

[1] Note that in Eq. (4.1), the cutoff magnitude M_0 is for the ETAS calculation, and here the cutoff magnitude Mc is based on the monitoring capability of a seismic network. Generally M_0 has to be higher than or equal to Mc. But in some cases, assuming that the above mentioned concept is clear, the differentiation of these two cutoff magnitudes is not purposely specified.

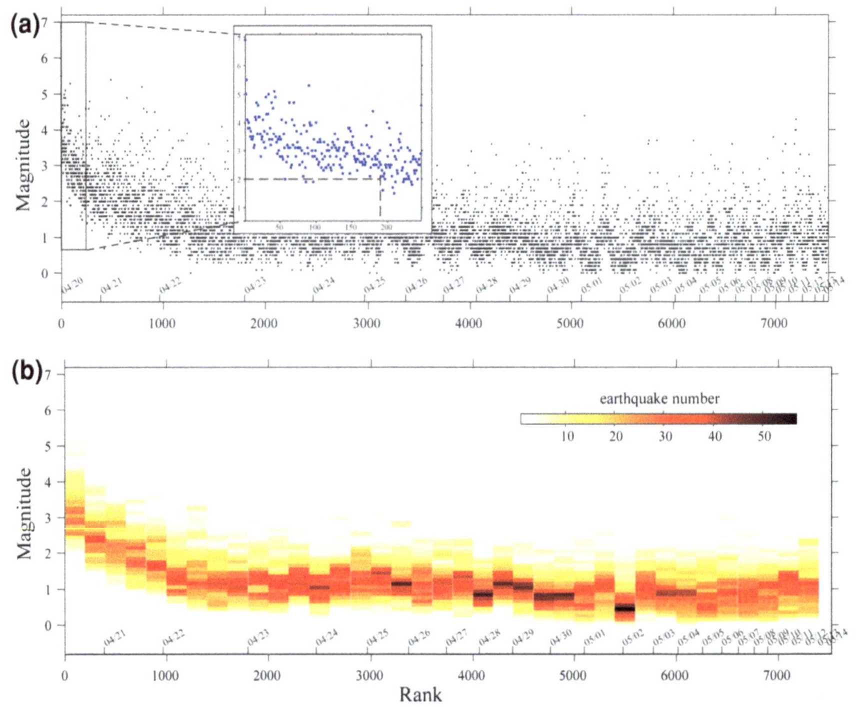

Fig. 4.6 Catalogue completeness of the Lushan M_S7.0 earthquake sequence. **a** Magnitude-rank distribution of the earthquake sequence. *Grey* subplot shows the partial enlarged view of the first 300 events, and the *vertical dashed line* indicates the completeness of M_L2.0 aftershocks, being the 190th event at 0.31 day after the mainshock; **b** Recorded number of earthquakes versus rank order

4.5 Summary and Discussion

Monitoring of the aftershock activities by the end of November 2013 obtains the following parameters:

- Maximum magnitude of the aftershock: M_S5.4
- b-value of the aftershock sequence: 0.72
- p-value and α-value in the ETAS model: 1.22 and 1.89, respectively
- length and width of the aftershock swarm: ~45 km and ~20 km, respectively
- focal depth ranges of the aftershock swarm: ~10 km to ~20 km
- strike of the planar main aftershock cluster: ~NE33°
- dip angle of the planar main aftershock cluster: ~41°

The above parameters can be compared with those of the mainshock: epicenter location relative to the centroid of the aftershock swarm ~NE60°, with distance ~5 km, and focal mechanism strike 15° ~ 40°, dip 43° ~ 59°, and slip 71° ~ 102°, according to the centroid moment tensor results of different agencies.

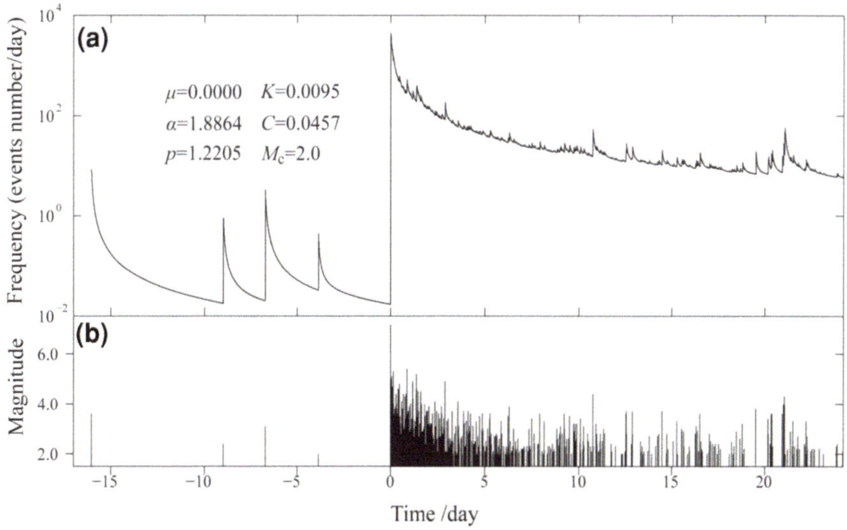

Fig. 4.7 Temporal variation of the conditional intensity by fitting the ETAS model to the Lushan M_S7.0 earthquake sequence, with cutoff magnitude M_L2.0. **a** Conditional intensity curve; **b** The magnitude-time plot as a reference

Fig. 4.8 Relationship between *p* and *b* values of aftershocks (reproduced from Wang (1994)), with the comparison with the Lushan sequence. The Lushan earthquake, marked by the *red square*, was not involved in the regression analysis. The regression line with slope −0.58 is denoted by a *solid line*, and the *dashed line* has a slope of −1

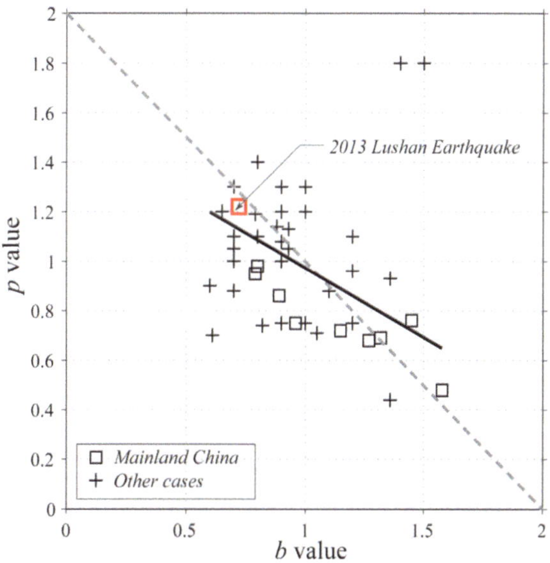

Although the observation and analysis of aftershock activity has almost been one of the routine works in seismological observation and interpretation, and there have been well formulations related to the properties of aftershocks such as the ETAS model and/or the CFS modeling, there are still some aspects in the aftershock

Table 4.1 Parameters of ETAS model with different cutoff magnitude

Cutoff magnitude	Total number/fitted number	μ	K	α	C	p
2.0	1,216/1,029	0.0000 ± 0.0000	0.0095 ± 0.0053	1.8864 ± 0.1856	0.0457 ± 0.0271	1.2205 ± 0.0712
2.1	1,068/883	0.0000 ± 0.0000	0.0084 ± 0.0053	1.9274 ± 0.1976	0.0335 ± 0.0194	1.2195 ± 0.0655
2.2	940/757	0.0000 ± 0.0000	0.0064 ± 0.0037	2.0280 ± 0.1835	0.0232 ± 0.0112	1.2245 ± 0.0511
2.3	832/651	0.0000 ± 0.0000	0.0052 ± 0.0034	2.0973 ± 0.1986	0.0208 ± 0.0105	1.2286 ± 0.0501
2.4	722/544	0.0000 ± 0.0000	0.0049 ± 0.0038	2.1287 ± 0.2388	0.0206 ± 0.0115	1.2437 ± 0.0545
2.5	626/458	0.0000 ± 0.0000	0.0030 ± 0.0023	2.2905 ± 0.2329	0.0148 ± 0.0083	1.2419 ± 0.0487
2.6	553/396	0.0000 ± 0.0000	0.0031 ± 0.0025	2.3030 ± 0.2413	0.0125 ± 0.0071	1.2519 ± 0.0512
2.7	492/342	0.1774 ± 0.8519	0.0039 ± 0.0033	2.2588 ± 0.2639	0.0144 ± 0.0094	1.2579 ± 0.0945
2.8	430/291	0.2265 ± 0.8403	0.0043 ± 0.0037	2.2387 ± 0.2818	0.0117 ± 0.0084	1.2336 ± 0.1018
2.9	360/240	0.0000 ± 0.0000	0.0032 ± 0.0035	2.3519 ± 0.3409	0.0107 ± 0.0093	1.2222 ± 0.0628
3.0	321/206	0.0000 ± 0.0000	0.0026 ± 0.0027	2.4456 ± 0.3213	0.0084 ± 0.0071	1.2221 ± 0.0628

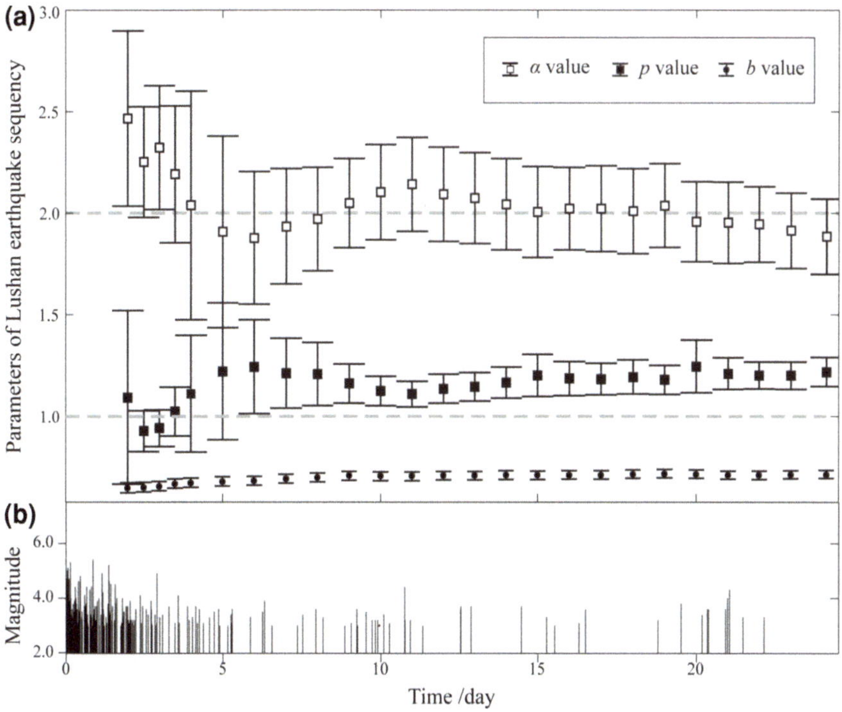

Fig. 4.9 Parameters α, p and b against the ending time in the ETAS fitting (**a**) and the magnitude-time plot (**b**)

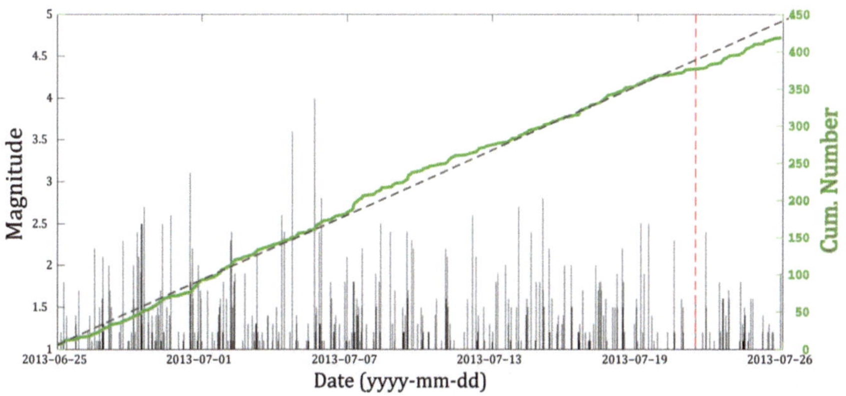

Fig. 4.10 Possible response of the Lushan aftershock sequence to the July 22, 2013 Minxian-Zhangxian, Gansu, M_S6.6 earthquake (the *vertical dashed line*, with epicentral distance about 485 km)

sequences which we do not fully understand. Figure 4.10 shows an example of such aspects related to the Lushan earthquake, that it seems that the sequence can respond to remote earthquakes. As shown in figure, a M_S6.6 earthquake 485 km away seems to have apparent impact on the Lushan sequence.

References

Fang LH, Wu JP, Zhang TZ, Huang J, Wang CZ, Yang T (2011) Yunnan Yingjiang Ms5.8 earthquake and its aftershock sequence relocations. Acta Seismol Sin 33:262–267 (in Chinese with English abstract)

Fang LH, Wu JP, Wang WL, Lü ZY, Wang CZ, Yang T, Cai Y (2013) Relocation of mainshock and aftershock sequences of M_S7.0 Sichuan Lushan earthquake. Chin Sci Bull 58:3451–3459. doi:10.1007/s11434-013-6000-2

Gao Y, Wang Q, Zhao B, Shi YT (2013) A rupture blank zone in middle south part of Longmenshan Faults: Effect after Lushan M_S7.0 earthquake of 20 April 2013 in Sichuan, China. Sci China: Earth Sci 43:1038–1046. doi:10.1007/s11430-013-4646-x (in Chinese)

Guo Z, Ogata Y (1997) Statistical relations between the parameters of aftershocks in time, space and magnitude. J Geophys Res 102:2857–2873. doi:10.1029/96JB02946

Han LB, Zeng XF, Jiang CS, Ni SD, Zhang HJ, Long F (2014) Focal mechanisms of the 2013 M_W6.6 Lushan, China earthquake and high-resolution aftershock relocations. Seism Res Lett 85:8–14. doi:10.1785/0220130083

Jiang CS, Zhuang JC, Long F, Han LB, Guo LJ (2013) Statistical analysis of ETAS parameters in the early stage of the 2013 Lushan M_S7.0 earthquake sequence. Acta Seismol Sin 35:661–669. doi:10.3969/j.jssn.0253-3782.2013.05.005 (in Chinese with English abstract)

Jiang HK, Zheng JC, Wu Q, Qu YJ, Li YL (2007) Earlier statistical feature of ETAS model parameters and their seismological meanings. Chin J Geophys 50:1778–1786 (in Chinese with English abstract)

Li ZW, Tian BF, Liu S, Yang JS (2013) Asperity of the 2013 Lushan earthquake in the eastern margin of Tibetan Plateau from seismic tomography and aftershock relocation. Geophys J Int 195:2016–2022. doi:10.1093/gji/ggt370

Liu CL, Zheng Y, Ge C, Xiong X, Xu HZ (2013) Rupture process of the M7.0 Lushan earthquake, 2013. Sci China: Earth Sci 56:1187–1192. doi:10.1007/s11430-013-4639-9

Lü J, Wang XS, Su JR, Pan LS, Li Z, Yin LW, Zeng XF, Deng H (2013) Hypocentral location and source mechanism of M_S7.0 Lushan earthquake sequence. Chin J Geophys 56:1753–1763. doi:10.6038/cjg20130533 (in Chinese with English abstract)

Miao M, Zhu SB (2013) The static Coulomb stress change of the 2013 Lushan M_S7.0 earthquake and its impact on the spatial distribution of aftershocks. Acta Seismol Sin 35:619–631. doi:10.3969/j.issn.0253-3782.2013.05.001 (in Chinese with English abstract)

Ogata Y (1989) Statistical model for standard seismicity and detection of anomalies by residual analysis. Tectonophysics 169:159–174. doi:10.1016/0040-1951(89)90191-1

Ogata Y (1992) Detection of precursory relative quiescence before great earthquakes through a statistical model. J Geophys Res 97:19845–19871. doi:10.1029/92JB00708

Omori F (1894) On aftershocks of earthquakes. J Coll Sci Imp Univ Tokyo 7:11–200

Utsu T (1961) A statistical study of on the occurrence of aftershocks. Geophys Mag 30:521–605

Waldhauser F, Ellsworth WL (2000) A double-difference earthquake location algorithm: method and application to the Northern Hayward Fault, California. Bull Seismol Soc Am 90:1353–1368

Wang J-H (1994) On the correlation of observed Gutenberg-Richter's b value and Omori's p value for aftershocks. Bull Seism Soc Am 84:2008–2011

Wang WM, Hao JL, Yao ZX (2013) Preliminary result for rupture process of April 20, 2013, Lushan earthquake, Sichuan, China. Chin J Geophys 56:1412–1417. doi:10.6038/cjg20130436 (in Chinese with English abstract)

Wu JP, Huang Y, Zhang TZ, Ming YH, Fang LH (2009) Aftershock distribution of the M_S8.0 Wenchuan earthquake and three dimensional P-wave velocity structure in and around source region. Chin J Geophys 52:320–328 (in Chinese with English abstract)

Xie ZJ, Jin BK, Zheng Y, Ge C, Xiong X, Xiong C, Xu HZ (2013) Source parameters inversion of the 2013 Lushan earthquake by combining teleseismic waveforms and local seismograms. Sci China: Earth Sci 56:1177–1186. doi:10.1007/s11430-013-4640-3

Zeng XF, Han LB, Shi YL (2013a) Focal mechanism solution of M_S7.0 Lushan earthquake, April 20, 2013. Sci Technol Rev 31(12):15–18. doi:10.3981/j.issn.1000-7857.2013.12.001 (in Chinese with English abstract)

Zeng XF, Luo Y, Han LB, Shi YL (2013b) The Lushan Ms7.0 earthquake on 20 April 2013: a high-angle thrust event. Chin J Geophys 56:1418–1424. doi:10.6038/cjg20130437 (in Chinese with English abstract)

Zhang GW, Lei JS (2013) Relocations of Lushan, Sichuan strong earthquake (M_S)7.0 and its aftershocks. Chin J Geophys 56:1764–1771. doi:10.6038/cjg20130534 (in Chinese with English abstract)

Zhang Y, Xu LS, Chen YT (2013) Rupture of the Lushan 4.20 earthquake and preliminary analysis on the disaster-causing mechanism. Chin J Geophys 56:1408–1411. doi:10.6038/cjg20130435 (in Chinese with English abstract)

Zhao B, Gao Y, Huang ZB, Jiang X, Li DH (2013a) Double difference relocation, focal mechanism and stress inversion of Lushan M_S7.0 earthquake sequence. Chin J Geophys 56:3385–3395. doi:10.6038/cjg20131014 (in Chinese with English abstract)

Zhao CP, Zhou LQ, Chen ZL (2013b) Source rupture process of Lushan M_S7.0 earthquake, Sichuan, China and its tectonic implication. Chin Sci Bull 58:3444–3450. doi:10.1007/s11434-013-6017-6

Chapter 5
An Aftershock of the 2008 Wenchuan Earthquake?

Abstract Whether the Lushan earthquake is one of the late aftershocks of the Wenchuan earthquake is a scientific issue subject to debate. This is by no means of only academic interest: In the perspective of seismic hazard assessment, if the Lushan earthquake can be regarded as one of the 'late' aftershocks of the Wenchuan earthquake, then the time-dependent seismic hazard of the southern Longmenshan fault zone has to be considered at the time scale of aftershock attenuation. Otherwise if the Lushan earthquake is an independent major earthquake, then the seismic hazard of the southern Longmenshan fault zone can be considered at the time scale of strong earthquake recurrence.

Keywords The 2013 Lushan earthquake · The 2008 Wenchuan earthquake · Aftershock · The Longmenshan fault zone · The ETAS model

5.1 Introduction: Pros and Cons Regarding to Whether the Lushan Earthquake is One of the Late Aftershocks of the Wenchuan Earthquake

The Lushan earthquake occurred nearly 5 years after the 2008 Wenchuan earthquake, located near to the epicenter of the Wenchuan earthquake, and shared the Longmenshan fault zone with the Wenchuan earthquake. Relation between the Lushan earthquake and the Wenchuan earthquake thus became one of the scientific issues subject to debate.

This chapter is written by Fang LH, Jiang CS, Jiang H, Zhang SF and Wu ZL (Institute of Geophysics, CEA).

Z. Wu et al., *Earthquake Phenomenology from the Field*, SpringerBriefs in Earth Sciences, DOI: 10.1007/978-981-4585-15-6_5,

5.1.1 The Lushan Earthquake is an Aftershock of the Wenchuan Earthquake

Wang et al. (2013) proposed that the Lushan earthquake can be considered as a delayed strong aftershock of the Wenchuan earthquake. Evidences for this argument include that the thrusting mechanism of the Lushan earthquake is similar to that of the Wenchuan earthquake, and the Lushan earthquake rupture occurred on the southern Longmenshan fault which was in the region with an increase of Coulomb failure stress (CFS) caused by the Wenchuan earthquake.

Chen et al. (2013a, b) pointed out that the argument that the Lushan earthquake is the largest aftershock, up to now, of the Wenchuan earthquake sequence, is more convincing, taking into account the hypocentral location, focal mechanism, magnitude and ruptured area of the Lushan earthquake and the Wenchuan earthquake. Bath's law that the strongest aftershock is generally 1.2 ($\pm$0.3) magnitude unit less than the mainshock was cited as a basis to support the aftershock argument (Li 2013). Chen et al. (2013a) also related the Lushan earthquake to the gap of seismic moment release of the Wenchuan earthquake inverted by seismic waveforms. As a matter of fact, in 2008, based on this gap of moment release, they forecasted that the gap could be filled in by a strong aftershock with magnitude M_W6.7–7.3 (see Fig. 2 of Chen et al. 2013a).

5.1.2 The Lushan Earthquake is Not an Aftershock of the Wenchuan Earthquake

Du et al. (2013) and Liu et al. (2013b) compared the focal mechanism, rupture process, aftershock distribution and surface rupture of the Lushan earthquake and the Wenchuan earthquake, and argued that the two earthquakes are significantly different. They stressed that the distance between the (epicenters of the) two earthquakes is about 90 km, and the distance between the two aftershock sequences is about 45 km; The rupture of the Wenchuan earthquake distributed mainly in the middle-to-northern segment of the Longmenshan fault zone, while the Lushan earthquake was located in the southern segment of the Longmenshan fault zone. They thus argued that the Lushan earthquake is not an aftershock of the Wenchuan earthquake.

5.1.3 The Lushan Earthquake is a 'New' Mainshock

The point of view that the Lushan earthquake was 'another shock' independent of the Wenchuan earthquake was broadcast by the CCTV News on April 20, 2013, through the interview with the seismologists at the China Earthquake Networks Center (CENC). This stimulated, to some extent, the media coverage of this scientific discussion. Spokesmen and experts of other seismological institutions, as well as experts publishing their views through the blog,[1] also joined in the discussion (Li 2013).

There are four reasons, summarizing the results of Du et al. (2013), Liu et al. (2013b), Xu et al. (2013a, b), and Zhang et al. (2013), to regard the Lushan earthquake as a new mainshock which is independent of the Wenchuan earthquake (Li 2013), although the Wenchuan earthquake may have some triggering effect on the Lushan earthquake: (1) the 45 km gap between the Wenchuan sequence and the Lushan sequence, and the location of the Lushan earthquake along a different fault than the central fault which accommodated the Wenchuan earthquake; (2) the Wenchuan sequence satisfying the Omori's law, which will be disrupted if the Lushan earthquake joined in the sequence; (3) the south-westward propagation of the Wenchuan earthquake rupture which was stopped by a strong barrier, to the other side of which is the Lushan earthquake; and (4) the Lushan sequence itself which satisfies the Omori's law, being seemingly independent of the Wenchuan sequence.

5.1.4 The Lushan Earthquake is Not a New Mainshock

Chen et al. (2013b) published a systematic comment on the above four reasons to regard the Lushan earthquake as an independent mainshock. They pointed out that while they were not in the position to insist on the aftershock argument, they did not agree with the mainshock argument either, that is, they thought that it is still too early to conclude that the Wenchuan earthquake sequence is a mainshock-aftershock sequence. As discussed by Chen et al. (2013b), due to the fact that the identification of an earthquake sequence is phenomenological and descriptive, with arbitrariness to some extent, whether the Lushan earthquake is a new mainshock or one of the aftershocks of the 2008 Wenchuan earthquake is based on empirical knowledge and will remain unsolved until the whole earthquake sequence has been completed.

5.2 The 2008 Wenchuan Earthquake, and Its Aftershocks

5.2.1 The May 12, 2008, Wenchuan Earthquake

For discussing the relation between the Lushan earthquake and the Wenchuan earthquake, one has to go back to the Wenchuan earthquake. The May 12, 2008, Wenchuan M_S8.0 earthquake is one of the great earthquakes, and the most devastating one, to have occurred in China since the beginning of the 21st century

[1] See: http://blog.sciencenet.cn/blog-51597-682325.html

(Chen and Booth 2011). Official data on August 25, 2008, from the central government of China indicated that the earthquake disaster caused 69,226 dead and 374,643 injured, plus 17,923 missing, with a significant portion of the loss of life due to the quake-induced landslides and/or rock falls. Direct economic loss was estimated as 845,100 million RMB Yuan.[2] These numbers of loss may explain why the public were so interested in the relation between the Lushan earthquake and the Wenchuan earthquake, even if 5 years had passed since the earthquake in 2008.

Field investigation and seismological/geological studies[3] revealed that the Wenchuan earthquake occurred along the Longmenshan fault zone, rupturing two parallel faults with about 240 and 72 km length, respectively, and causing intensity up to XI on the Modified Mercalli Intensity (MMI) scale, with peak ground acceleration (PGA) close to 1 g at the closest-epicenter-distance strong-motion recording station (~23 km). The narrow-ellipsoid-shaped MMI-IX isoseismal is almost collocated with the rupture zone, being about 300 km long striking north-eastward from Dujiangyan to Guangyuan, Sichuan Province. The epicenter location, or the surface projection of the rupture initiation point leading the commence of this great earthquake, was determined to be 30°57′N, 103°24′E. The centroid of the rupture located about 200 km north-eastward, clearly indicating the rupture propagation process. The focal mechanism of this earthquake was shown to be thrust-predominant, with the dip angle of the earthquake fault about 60°. With noticeable uncertainties, the initiation point, as indicated by the hypocenter, was determined by near-source seismic recordings to be about 15 km, while the rupture spanned from the surface of the ground to the lower crust about 30 km deep. The whole rupture process can be divided into several stages, corresponding to a series of sub-events, with two main asperities (located in Wenchuan and Beichuan, respectively) broken during the earthquake. Rupture duration was determined as about 105 s, with the stage of 12–45 s containing about 50 % of the whole energy released by seismic waves.

5.2.2 The Aftershocks of the Wenchuan Earthquake

Table 5.1 and Fig. 5.1 show the aftershocks of the Wenchuan sequence. The whole aftershock zone is basically consistent with the rupture zone of the mainshock, which can be divided into the northern and the southern part. Estimates of the duration, maximum magnitude, rate, and most-likely locations, of the strong aftershocks, based on the presently available knowledge of seismicity, such as the Omori's law and the Gutenberg-Richter's law, were shown to be consistent with the observed aftershocks. Scientists of the IUGG Commission on Geophysical

[2] In 2008, the foreign exchange rate was 1 USD ≈ 7 RMB Yuan. April 2008 was the crossing point when the rate changed from above 7 to below 7.

[3] See: http://www.journals.elsevier.com/tectonophysics/virtual-special-issues/virtual-special-issue-on-the-2008-wenchuan-earthquake/; and http://www.wceq.org/.

Table 5.1 The Wenchuan earthquake sequence (above M_S5.5) data from the China Earthquake Networks Center (CENC)

No.	Date	Origin time (Local)	Lat.	Lon.	M_S
1	2008-5-12	14:28:04	31.0	103.4	8.0 (mainshock)
2	2008-5-12	14:36:39	31.3	103.6	5.8
3	2008-5-12	14:43:14	31.3	103.7	6.0
4	2008-5-12	19:11:00	31.3	103.4	6.0
5	2008-5-13	04:08:49	31.4	103.8	5.6
6	2008-5-13	15:07:07	31.0	103.2	6.1
7	2008-5-14	10:54:36	31.3	103.4	5.6
8	2008-5-16	13:25:46	31.4	103.2	5.9
9	2008-5-18	01:08:25	32.3	104.9	6.0
10	2008-5-25	16:21:49	32.6	105.3	6.4 (the largest aftershock)
11	2008-5-27	16:37:51	32.8	105.6	5.7
12	2008-7-24	03:54:44	32.8	105.5	5.6
13	2008-7-24	15:09:29	32.8	105.5	6.0
14	2008-8-01	16:32:42	32.1	104.7	6.1
15	2008-8-05	17:49:16	32.8	105.5	6.1
16	2008-9-12	01:38:59	33.0	105.6	5.7
17	2009-6-30	02:03:50	31.5	104.0	5.8
18	2011-11-1	05:58:15	32.6	105.2	5.6

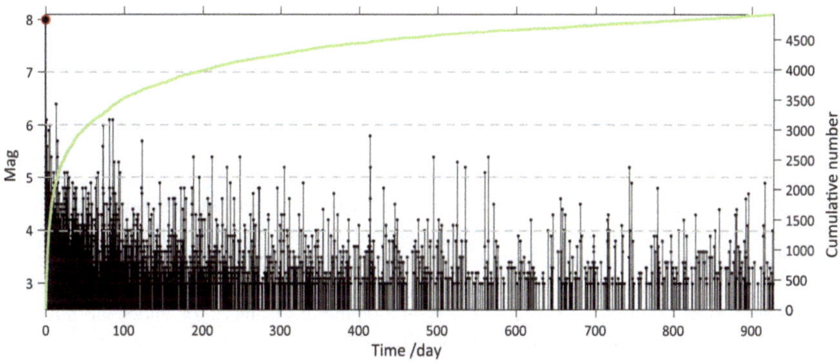

Fig. 5.1 Aftershock sequence of the 2008 Wenchuan earthquake, up to November 28, 2010, about 900 days after the mainshock

Risk and Sustainability (GRC) and the US Geological Survey (USGS) were invited to join in the predictive study of the aftershock tendency.[4] This estimate helped much to the rescue and reconstruction actions, which was especially noticeable considering the landslides and landslide lakes threatening the relief and reconstruction. Remarkably, on May 19, 2008, a false alarm of a 'magnitude 7 aftershock' broadcast on local TV caused widespread social disorder in Chengdu,

[4] See Appendix 2 of this Brief.

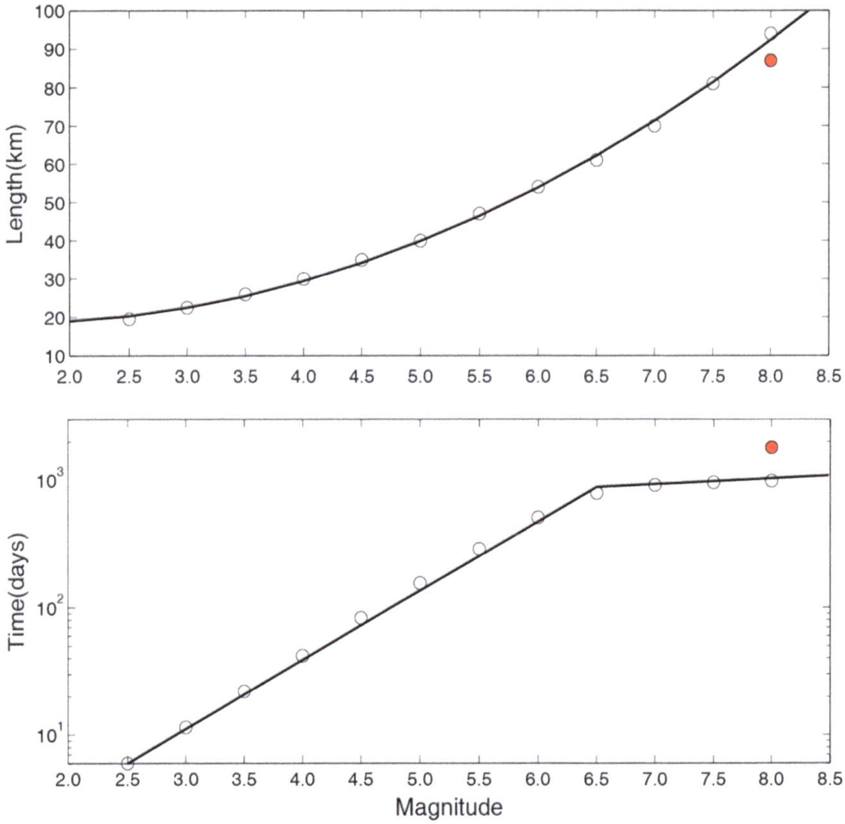

Fig. 5.2 Gardner and Knopoff (1974) definition of aftershocks and the relation between the Lushan earthquake and the Wenchuan earthquake (*red dot*). In the figure, the *horizontal axis* shows the magnitude of the mainshock (for the Wenchuan-Lushan relation, 8.0), and *vertical axes* show the spatial and temporal distances, respectively, between an aftershock and the mainshock. Those events within the range of the spatial-temporal distances, or under the *threshold line* shown in the figure, may be regarded as an 'aftershock'

Xi'an, and other cities near the Wenchuan earthquake. This can partly explain why in the emergency response stage of the 2013 Lushan earthquake, the public were still sensitive to the wording of 'the aftershock of the Wenchuan earthquake'.

The debate about the relation between the Lushan earthquake and the Wenchuan earthquake was probably further highlighted by two cultural factors. One may be that the scientific discussion was paid attention to by the news media (Li 2013); The other might be that the Chinese translation of the word 'aftershock' in the Chinese mainland (Yu-Zhen) indicates that aftershocks have physical relations with the mainshock, that Yu means 'remained' and Zhen means 'quake', which is different from the Chinese translation in other places such as Taiwan (Hou-Zhen) that Hou means simply 'after'.

But scientifically there did exist some tricky issues to be considered. Figure 5.2 shows the 'aftershock paradox' using the empirical relation of Gardner and

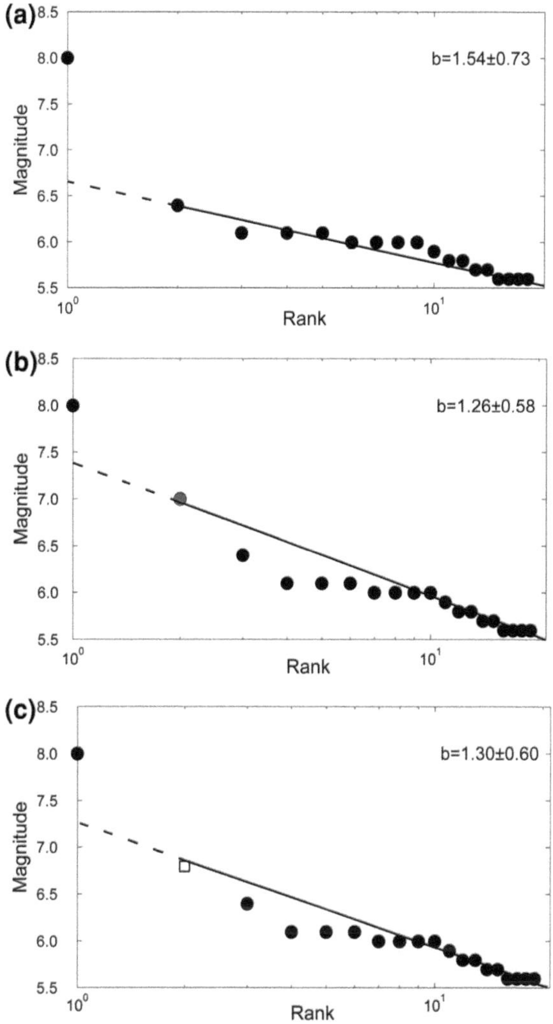

Fig. 5.3 Rank-ordering plot of the Wenchuan aftershock sequence, considering the role of the sub-rupture (the *open square*) and the Lushan earthquake (the *red dot*). Rank-ordering plot is a validation of the well-known Zipf-distribution for identify the power-law-like scaling in the frequency-size distribution from under-sampled data (Sornette et al. 1996), which orders by rank the quantity to be studied by placing the largest in the first rank, the second largest in the second rank, and so on, and then plots the quantity versus their rank order on a log-log plot. If there exists a frequency-size distribution of a power-law type, then the rank-ordering analysis can show the existence of the power-law by a straight line on the double-log rank-ordering plot. In the figure, the most-likelihood estimate of the *b*-value, together with its uncertainty, is also given. See text for details

Knopoff (1974) to define an aftershock. From the figure it may be seen that the relation between the Wenchuan earthquake and the Lushan earthquake is to much extent ‘marginal’, that considering the spatial distance the Lushan earthquake can be considered as an aftershock, but considering the temporal span the Lushan earthquake can not be considered as an aftershock.

Figure 5.3 shows another perspective that considering the aftershock activity of the Wenchuan earthquake the sequence seems too ‘abnormal’ that the magnitude difference between the mainshock and the largest aftershock is too large, and the *b*-value of the aftershock sequence is too large, being 1.5. If the Lushan earthquake is included in this aftershock sequence, as shown in Fig. 5.3b, then a smaller magnitude difference from the mainshock and a more reasonable *b*-value can be obtained. It seems, therefore, that the Lushan earthquake can be regarded as one of the aftershocks of the Wenchuan sequence. On the other hand, however, it has to be noticed that accompanied with the main rupture of the Wenchuan earthquake which is along the central fault, there was a sub-rupture along the front-range fault, with the surface rupture being ~70 km long. If this sub-rupture is considered as an 'immediate aftershock', then as shown in Fig. 5.3c, one can still obtain a smaller difference between the magnitude of the largest aftershock and the mainshock, together with a more reasonable *b*-value.

5.3 The Relationship Between the Lushan Earthquake and the Wenchuan Earthquake from a Statistical Viewpoint

5.3.1 Space-Time Epidemic Type Aftershock Sequence Model and Stochastic Declustering Method

Intrinsically it is difficult to distinguish whether an earthquake is a 'background event' or if it is triggered by another: A quantitative physical definition, therefore, has never been given. That is, taken arbitrarily an earthquake, seen only with the source parameters of this event, such as the stress drop or source duration, it is hard to conclude whether it is an aftershock of another one, or an independent mainshock.

To cope with this complicated situation, the space-time Epidemic Type Aftershock Sequence (ETAS) model introduced by Ogata (1998) combines the Omori-Utsu law, productivity of aftershocks, and the Gutenberg-Richter law, providing a probabilistic estimation for the declustering. In the space-time ETAS model, the expected number of earthquakes above magnitude threshold M_c in a unit space-time window centered at time t and spatial location (x, y), can be expressed by

$$\lambda(t,x,y) = \mu(x,y) + \sum_{i:t_i<t} \kappa(m_i)g(t-t_i)f(x-x_i, y-y_i; m_i). \tag{5.1}$$

In Eq. (5.1), $\mu(x, y)$ represents the background seismicity rate, $\kappa(m)$ is the expected number of aftershocks generated from an event of magnitude m; and $g(t)$ and $f(x, y; m)$ are the probability density function of the time interval distribution and the relative locations between aftershocks and the mainshock, respectively, being represented as:

$$\kappa(m) = Ae^{\alpha(m-m_c)}, \quad m \geq m_c, \tag{5.2}$$

$$g(t) = \frac{p-1}{c}\left(1+\frac{t}{c}\right)^{-p}, \quad t > 0, \tag{5.3}$$

$$f(x,y;m) = \frac{q-1}{\pi De^{\gamma(m-m_c)}}\left(1+\frac{x^2+y^2}{De^{\gamma(m-m_c)}}\right)^{-q}. \tag{5.4}$$

The parameters A, α, c, p, D, q and γ can be estimated through the maximum likelihood method. As a validation of the above mentioned principle, the stochastic declustering method proposed by Zhuang et al. (2002) obtains the background probability of each event. In the algorithm, each event gets a probability of being either a 'background event' or a direct offspring triggered by others, and the probability of the jth event as a 'background event' is:

$$\varphi_j = \frac{\mu(x_j, y_j)}{\lambda(t_j, x_j, y_j)}. \tag{5.5}$$

5.3.2 *Probability for the Lushan Earthquake as an Aftershock, and Contribution of the Wenchuan Earthquake to the Occurrence of the Lushan Earthquake*

Using the space-time ETAS model and the stochastic declustering method, Jia et al. (2014) analyzed the earthquakes in the Lushan region to discuss the relationship between the Wenchuan earthquake and the Lushan earthquake. In the calculation, one of the important improvements of the algorithm is that the Wenchuan earthquake cannot be considered simply as a 'point source': Due to the limited range of the earthquake rupture (the surface fault of the main rupture spanned 240 km, with two main asperities located in Wenchuan and Beichuan, respectively), in the ETAS calculation, the Wenchuan earthquake should be treated as 'a series of major earthquakes'.

By the ETAS analysis, Jia et al. (2014) obtained that the probability of the Lushan earthquake as a background event is 88 %, and the probability for it to be an aftershock is 12 %. They also found that the Wenchuan earthquake changed the local seismicity in the Lushan region when the cumulative background probability (Zhuang et al. 2005) of earthquakes is considered. The indirect triggering probability is 50 % when the background probability can be divided into the indirect triggering of the Wenchuan earthquake and the original background seismicity. They thus concluded that the contribution of the Wenchuan earthquake to the occurrence of the Lushan earthquake is 62 % (50 + 12 %), and the remaining 38 % is the contribution of other previous earthquakes.

5.4 Discussion without Conclusion

Calculation of the stress triggering effect of the Wenchuan earthquake on the Lushan earthquake indicates that there was close relation between these two earthquakes (Lei et al. 2013; Liu et al. 2013a; Shan et al. 2013; Wu et al. 2013; Parsons and Segou 2014; Wang et al. 2014). But how to define the relation between the Lushan earthquake and the Wenchuan earthquake, especially whether the Lushan earthquake can be regarded as one of the 'late aftershocks' of the Wenchuan earthquake (Chen et al. 2013a, b; Du et al. 2013; Liu et al. 2013b), is one of the questions subject to debate.

Whether an earthquake can be associated with the aftershock sequence of another earthquake is one of the tricky questions in earthquake seismology, although clues can be obtained from the (somehow arbitrary) definition of aftershocks (Knopoff and Gardner 1972; Gardner and Knopoff 1974; Knopoff 2000), scaling of aftershock zone (Kisslinger and Jones 1991; Kagan 2002), CFS variation (Parsons 2002), Omori-law-type temporal decay (Utsu et al. 1995; Yang and Ben-Zion 2009; Ziv 2006), and relation with the mainshock rupture (Das and Henry 2003; Kagan and Houston 2005; Woessner et al. 2006; Wong and Schoenberg 2009). The discussion in the perspective of ETAS model (Jia et al. 2014) can provide some quantitative indexes describing the relation between an earthquake and another.

For the Lushan earthquake, its relation with the Wenchuan earthquake seems not as simple as one expected. The comment of Chen et al. (2013b) is correct that at the present time, much more important than the debate on whether the Lushan earthquake can be regarded as one of the aftershocks of the Wenchuan earthquake, is to seriously consider the seismic hazard in the southern Longmenshan fault zone. However, considering the two 'seismic gaps' left after the Lushan earthquake (Gao et al. 2013), the debate itself has direct implication for the seismic hazard: If the Lushan earthquake can be regarded as one of the 'late' aftershocks of the Wenchuan earthquake, then the time-dependent seismic hazard of the two 'seismic gaps' in the southern Longmenshan fault zone has to be considered at the time scale of aftershock attenuation; Otherwise if the Lushan earthquake is an independent major earthquake, then the seismic hazard of the southern Longmenshan fault zone can be considered at the time scale of strong earthquake recurrence. Retrospective case lesson of the Annual Consultation is somehow educative[5] (for the general introduction of the Annual Consultation, see Wu (1997) and Zhao et al. (2010)): In 2009 and 2010, the southern Longmensha fault zone was delineated as the seismically risky region in the year, because the expert panel were considering the problem with the time scale of aftershocks. In 2011, 2012, and 2013, however, the southern Longmenshan fault zone was out of the seismically risky region in the year, because experts turned their consideration to the time scale of strong earthquake recurrence.

References

Chen Y, Booth DC (2011) The Wenchuan earthquake of 2008: anatomy of a disaster. Science Press in Cooperation with Springer, Beijing

Chen YT, Yang ZX, Zhang Y, Liu C (2013a) From 2008 Wenchuan earthquake to 2013 Lushan earthquake. Scientia Sinica Terrae 43:1064–1072 (in Chinese)

Chen YT, Yang ZX, Zhang Y, Liu C (2013b) A brief talk on the 20 April 2013 Lushan M_W6.7 earthquake. Acta Seismol Sin 35:285–295. doi:10.3969/j.issn.0253-3782.2013.03.001 (in Chinese with English abstract)

[5] According to the internal report of the Annual Consultation, 2009, 2010, 2011, 2012, and 2013. Materials provided by the China Earthquake Networks Center (CENC).

Das S, Henry C (2003) Spatial relation between main earthquake slip and its aftershock distribution. Rev Geophys 41(3):1013. doi:10.1029/2002RG000119

Du F, Long F, Ruan X, Yi GX, Gong Y, Zhao M, Zhang ZW, Qiao HZ, Wang Z, Wu J (2013) The $M7.0$ earthquake and the relationship with the $M8.0$ Wenchuan earthquake in Sichuan, China. Chin J Geophys 56:1772–1783. doi:10.6038/cjg20130535 (in Chinese with English abstract)

Gardner JK, Knopoff L (1974) Is the sequence of earthquakes in Southern California, with aftershocks removed, Poissonian? Bull Seism Soc Am 64:1363–1367

Gao Y, Wang Q, Zhao B, Shi YT (2013) A rupture blank zone in middle south part of Longmenshan faults: effect after Lushan $M_S7.0$ earthquake of 20 April 2013 in Sichuan, China. Sci China: Earth Sci 43:1038–1046. doi:10.1007/s11430-013-4646-x (in Chinese)

Jia K, Zhou SY, Zhuang JC, Jiang CS (2014) Possibility of the independence between the 2013 Lushan earthquake and the 2008 Wenchuan earthquake on Longmen Shan fault, Sichuan. Seism Res Lett 85:60–67. doi:10.1785/0220130115

Kagan YY (2002) Aftershock zone scaling. Bull Seism Soc Am 92:641–655. doi:10.1785/0120010172

Kagan YY, Houston H (2005) Relation between mainshock rupture process and Omori' law for aftershock moment release rate. Geoghys J Int 163:1039–1048. doi:10.1111/j.1365-246x.2005.02772.x

Kisslinger C, Jones LM (1991) Properties of aftershock sequences in Southern California. J Geophys Res 96:11947–11958. doi:10.1029/91JB01200

Knopoff L (2000) The magnitude distribution of declustered earthquakes in Southern California. Proc Natl Acad Sci USA 97:11880–11884

Knopoff L, Gardner JK (1972) Higher seismic activity during local night on the raw worldwide earthquake catalogue. Geophys J R astr Soc 28:311–313. doi:10.1111/j.1365-246X.1972.tb06133.x

Lei XL, Ma SL, Su JR, Wang XL (2013) Inelastic triggering of the 2013 $M_W6.6$ Lushan earthquake by the 2008 $M_W7.9$ Wenchuan earthquake. Seismol Geol 35:411–422. doi:10.3969/j.issn.0253-4967.2013.02.019 (in Chinese with English abstract)

Li N (2013) What's the point of arguing about "whether Lushan earthquake is an aftershock of Wenchuan earthquake"? Sci Technol Rev 31(12):8–9 (in Chinese)

Liu BY, Shi BP, Lei JS (2013a) Effect of Wenchuan earthquake on probabilities of occurrence of Lushan and surrounding faults. Acta Seismol Sin 35:642–651. doi:10.3969/j.issn.0253-3782.2013.05.003 (in Chinese with English abstract)

Liu J, Yi GX, Zhang ZW, Guan ZJ, Ruan X, Long F, Du F (2013b) Introduction to the Lushan, Sichuan $M7.0$ earthquake on 20 April 2013. Chin J Geophys 56:1404–1407. doi:10.6038/cig20130434 (in Chinese with English abstract)

Ogata Y (1998) Space-time point-process models for earthquake occurrences. Ann Inst Stat Math 50:379–402. doi:10.1023/A:1003403601725

Parsons T (2002) Global Omori law decay of triggered earthquakes: large aftershocks outside the classical aftershock zone. J Geophys Res 107:2156–2202. doi:10.1029/2001JB000646

Parsons T, Segou M (2014) Stress, distance, magnitude, and clustering influences on the success or failure of an aftershock forecast: the 2013 $M = 6.6$ Lushan earthquake and other examples. Seism Res Lett 85:44–51. doi:10.1785/0220130100

Sornette D, Knopoff L, Kagan YY, Vanneste C (1996) Rank-ordering statistics of extreme events: application to the distribution of large earthquakes. J Geophys Res 101:13883–13893

Shan B, Xiong X, Zheng Y, Jin BK, Liu CL, Xie ZJ, Xu HZ (2013) Stress changes on major faults caused by 2013 Lushan earthquake, and its relationship with 2008 Wenchuan earthquake. Sci China: Earth Sci 43:11169–11176. doi:10.1007/s11430-013-4642-1

Utsu T, Ogata Y, Matsu'ura RS (1995) The centenary of the Omori formula for a decay law of aftershock activity. J Phys Earth 43:1–33

Wang WM, Hao JL, Yao ZX (2013) Preliminary result for rupture process of Apr. 20, 2013, Lushan Earthquake, Sichuan, China. Chin J Geophys 56:1412–1417. doi:10.6038/cjg20130436 (in Chinese with English abstract)

Wang Y, Wang F, Wang M, Shen Z-K, Wan Y (2014) Coulomb stress change and evolution induced by the 2008 Wenchuan earthquake and its delayed triggering of the 2013 M_W6.6 Lushan earthquake. Seism Res Lett 85:52–59. doi:10.1785/0220130111

Woessner J, Schorlemmer D, Wiemer S, Mai PM (2006) Spatial correlation of aftershock locations and on-fault main shock properties. J Geophys Res 111:B08301. doi:10.1029/2005JB003961

Wong K, Schoenberg FP (2009) On mainshock focal mechanisms and the spatial distribution of aftershocks. Bull Seism Soc Am 99:3402–3412. doi:10.1785/0120080371

Wu FT (1997) The annual earthquake prediction conference in China (National consultative meeting on seismic tendency). Pure appl Geophys 149:249–264

Wu JC, Yu S, Cai YJ, Lei DN, Li H (2013) Stress triggering of the Lushan M7.0 earthquake by the Wenchuan Ms8.0 earthquake. Geodesy Geodyn 4(3):35–39. doi:10.3724/SP.J.1246.2013.03035

Xu XW, Chen GH, Yu GH, Cheng J, Tan XB, Zhu AL, Wen XZ (2013a) Seismogenic structure of Lushan earthquake and its relationship with Wenchuan earthquake. Earth Sci Front 20(3):11–20 (in Chinese with English abstract)

Xu XW, Wen XZ, Han ZJ, Chen GH, Li CY, Zheng WJ, Zhang SM, Ren ZK, Xu C, Tan XB, Wei ZY, Wang MM, Ren JJ, He Z, Liang MJ (2013b) Lushan M_S7.0 earthquake: a blind reserve-fault earthquake. Chin Sci Bull 58:3437–3443. doi:10.1007/s11434-013-5999-4

Yang WZ, Ben-Zion Y (2009) Observational analysis of correlations between aftershock productivities and regional conditions in the context of a damage rheology model. Geophys J Int 177:481–490

Zhang YQ, Dong SW, Hou CT, Shi JS, Wu ZH, Li HL, Sun P, Liu G, Li J (2013) Seismogenic structure of the April 20, 2013, Lushan M_S7 earthquake in Sichuan. Acta Geol Sinica 87:633–645

Zhao YZ, Wu ZL, Jiang CS, Zhu CZ (2010) Reverse tracing of precursors applied to the annual earthquake forecast: retrospective test of the annual consultation in the Sichuan-Yunnan Region of southwest China. Pure appl Geophys 167:783–800. doi:10.1007/s00024-010-0077-1

Zhuang JC, Chang CP, Ogata Y, Chen YI (2005) A study on the background and clustering seismicity in the Taiwan region by using point process models. J Geophys Res 110:B05S18. doi:10.1029/2004JB003157

Zhuang JC, Ogata Y, Vere-Jones D (2002) Stochastic declustering of space-time earthquake occurrences. J Am Stat Assoc 97:369–380. doi:10.1198/016214502760046925

Ziv A (2006) Does aftershock duration scale with mainshock size? Geophys Res Lett 33(17):L17317. doi:10.1029/2006GL027141

Chapter 6
Scientific Products Serving the Earthquake Emergency

Abstract Serving the emergency response to the Lushan earthquake, several scientific products were produced and disseminated by the CEA institutions. This system of emergency-oriented scientific information service, which had been established since the 2008 Wenchuan earthquake, is tested by the earthquake emergency response. Quality control and improvement of such scientific products need a systems engineering perspective, which could be understood using the language of the technology readiness level (TRL).

Keywords Earthquake emergency · Seismological information service · Technology readiness level (TRL) · Public understanding

6.1 Introduction

On May 12, 2013, at the 5th anniversary of the Wenchuan earthquake, Department for Earthquake Monitoring and Prediction of China Earthquake Administration (CEA) published its Open File *Collection of the Visualized Emergency Service Products Produced by the Seismological Monitoring System of China: the April 20, 2013, Lushan, Sichuan, Major Earthquake.*[1] This atlas summarizes the seismological products serving the earthquake emergency response, which includes the fast determined location and magnitude of the mainshock; its focal mechanism

This chapter is written by Jiang CS and Wu ZL (Institute of Geophysics, CEA), assisted by Zhang SF and Jiang H, with thanks to the Working Group for Technical Supporting of Earthquake Emergency Response, Institute of Geophysics, CEA (group leaders Yang JS, Li XJ and Zhang DN) and the Department for Earthquake Monitoring and Prediction, CEA (directors Li K, Song YY and Che S), for the materials used in this chapter. We thank the US Geological Survey (USGS) which shared its experiences with CEA regarding to the seismological information service.

[1] Eds Yin CM, internal publication (in Chinese).

DOI: 10.1007/978-981-4585-15-6_6, © The Author(s) 2014

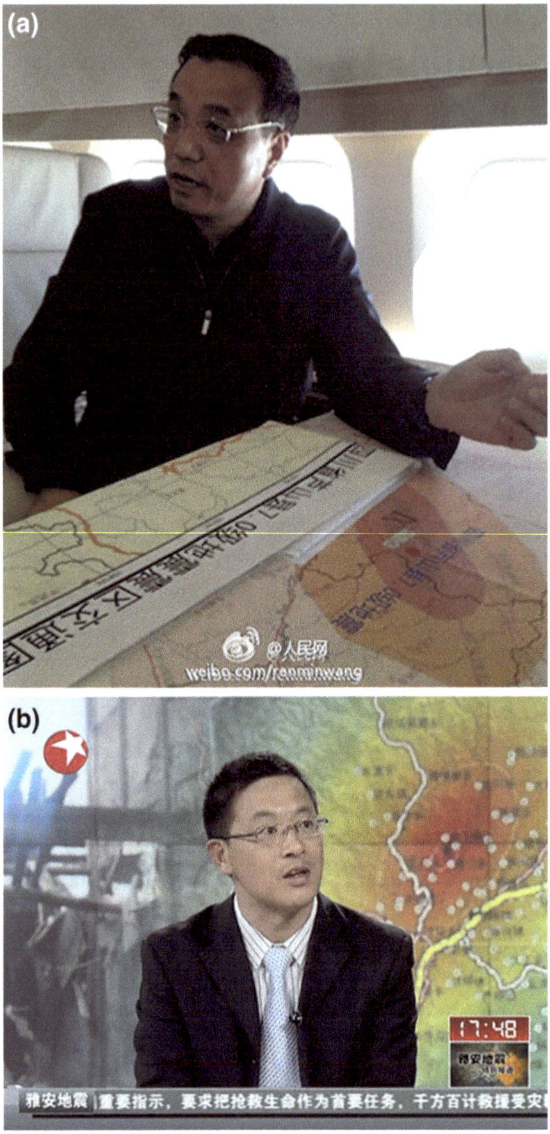

Fig. 6.1 **a** Chinese Premier Li Keqiang on the airplane to the epicentral region, with the maps of predicted intensity distribution and local transportation facilitating the commanding of the emergency rescue and relief deployment. *Picture* from the official news media (weibo.com/renminwang), April 20, 2013. **b** News focus program using ShakeMap as the background. Snapshot of TV program on April 20, courtesy of Zhang DN, Institute of Geophysics, CEA. The *Chinese characters* to the bottom shows the commands of the Chinese president to put life saving as the first priority

and rupture process; ShakeMap for prediction of intensity distribution based on the location, focal mechanism, and rupture process; and relocation and focal mechanism of the aftershocks, together with the distribution of seismic stations in the epicentral region, including the mobile seismic stations deployed for emergency field observation. Change of static Coulomb failure stress (CFS) caused by the mainshock, regional seismotectonic map, regional map of stress state, and focal mechanism of historical earthquakes in the surrounding regions are also included. Institutions contributing to the above mentioned products include the Institute of Geophysics, CEA;

the China Earthquake Networks Center (CENC); the Institute of Earthquake Science,[2] CEA; the Institute of Crustal Geodynamics,[3] CEA; and the Institute of Geology, CEA, with the assistance of provincial earthquake administrations.

Immediately after the earthquake, these scientific products were produced by the related institutions and sent to the CEA headquarter via electronic channel. Some of these products were reproduced by assimilating with related information such as transportation and population. The reproduced maps were sent to the central government for serving the commanding of the earthquake rescue and relief. On April 20, 2013, Chinese Premier Li Keqiang flew to the disastrous region. Some of the hands-on maps facilitating his decision-making on the airplane were based on these scientific products (see Fig. 6.1a). Some of the scientific products were also used by TV programs to communicate with the public (Fig. 6.1b).

6.2 Timeliness and Contents of the Scientific Products

6.2.1 Quickly Determined Parameters

For comparing with the refined results based on more data and more research, and the evaluation of the scientific products, the quickly determined parameters are briefly described as follow, mainly based on the Open File compiled by the Department for Earthquake Monitoring and Prediction, CEA.[4] Note that the parameters listed here were actually the 'final results of the quick report', that is, there had been already several rounds of revisions to the 'first-glance result', or in another word, they were the 'version X.0 results' of the quick report.

Location and magnitude of the mainshock was firstly broadcast by Microblogs[5] and then by cell phone message. The official quick report, that is, formal report to the central government for emergency decision-making, gave the epicenter location 30.3°N × 103.0°E, with origin time 08:02:46.0 (Beijing time), and focal depth 13 km. Uncertainty of the location was not provided, which is one of the aspects to be improved in future. Surface wave magnitude was determined as 7.0. This quick report was almost in fit with the final result. For this earthquake, however, the accurate location of the mainshock showed critical importance for understanding the seismotectonics (see Chap. 3), and the refined location acted not only as an improvement of the quick report but also as a necessary step in the investigation of the earthquake.

Focal mechanism of the mainshock, using teleseismic waveforms, regional and local waveforms, and regional P-wave first motion data, obtained a thrust dominated focal mechanism, being consistent with each other, with a NNE-striking fault, and

[2] Direct word-by-word translation from Chinese: Institute of Earthquake Forecast.

[3] Direct word-by-word translation from Chinese: Institute of Crustal Stress.

[4] See footnote 1.

[5] 'Micro-blog' is the Chinese version of Twitter. See Chap. 7 of this Brief.

a near 45° dip angle. Moment magnitudes measured were between 6.5 and 6.7, and centroid depth was determined as 12 km. Consistency among different determinations as per the focal mechanism was partly because that the rupture process of this event was relatively simple. Later after the earthquake, the related results were revised and published (Xu et al. 2013; Zhang et al. 2013; Zhao et al. 2013). Source time function (STF) of the earthquake rupture showed a two-stage character with the main rupture lasted for about 10 s, and a 'near-isotropic' circular rupture propagation (Zhang et al. 2013). There was a debate about the surface wave magnitude which is apparently larger than the moment magnitude, being inconsistent with the well-known concept of 'magnitude saturation'. One of the suspects was that the quick report may be problematic, which was resolved by the detailed investigation (Liu et al. 2013).

Firstly based on the location, magnitude, and site condition, and secondly based on the focal mechanism and rupture process, different versions of ShakeMap were produced which predicted the distribution of seismic intensity and strong ground motion. For the government and the public, this information is more directly relevant to the rescue and relief.

Aftershock distribution is another important constraint on the property of the mainshock. As one of the components of the emergency response, coordinated by the CEA, the Earthquake Administration of Sichuan Province, the Earthquake Administration of Chongqing Municipality, and the Institute of Geophysics, CEA, deployed a real-time on-site monitoring network consisting of 15 seismic stations. Joint data analysis with local and regional seismic stations using the double difference (DD) algorithm provided precise locations of aftershocks (see Chap. 4 of this Brief). Location result shows that the aftershock sequence could be approximated as a plane striking NNE and dipping NNW, with dip angle near 45°, and depth distribution between 10 and 20 km. Quick reports of the aftershock distribution also found some secondary-level features, which were revised or refined with more data accumulated. Refined results of the aftershocks were later published (Fang et al. 2013). Focal mechanisms of aftershocks above $M_L 4.0$ were determined using seismic waveform data, showing thrust dominated mechanisms, being similar to the mainshock.

6.2.2 Timeliness of the Scientific Products and On-going Modification

According to the regulation of the Department for Earthquake Monitoring and Prediction, CEA, and the Department for Earthquake Emergency and Rescue, CEA, the CENC is responsible for the quick report of the location and magnitude. After the quick report, the response process is automatically initiated. Focal mechanism of the mainshock is provided within 1 h, and rupture process is provided within 2 h, after the quick report.

ShakeMap based on location, magnitude and site condition is produced within 1 h, providing the maximum intensity and the affected area, and is further revised based on the focal mechanism and rupture process, providing detailed intensity distribution. In the regions with strong motion network deployment, ShakeMap is further revised and refined using the strong motion data collected. In parallel, using the empirical relation between magnitude, exposure and losses, fatalities and affected area are predicted, which is one of the important piece of information assisting the emergency decision-making.

Rupture process and focal mechanism of the mainshock are inverted by different institutions independently, using different datasets available and different algorithms. This is partly because, for many earthquakes, the results of rupture process by different approaches, especially at the quick report stage, are different to much extent. The comparison among different results acts both as a dynamo for the improvement of each approach and as a reference to the decision.

Based on the scientific products directly related to the monitoring, some of the 'deduced' results are also generated. One example is the change of the CFS caused by the mainshock, which acts both as an indicator of aftershock distribution and as an expectation of the change of earthquake probabilities on the active faults surrounding the epicenter.

Some of the emergency-oriented scientific results are database based, and are quickly provided after the earthquake. Examples of this kind of products include the information of historical earthquakes, seismogenic structures and stress state, and economic and social information such as transportation and population.

6.3 Maintenance and Quality Control of the Scientific Products

6.3.1 Development of the Information Service System

The May 12, 2008, Wenchuan earthquake marked a turning point of the seismological information service provided by the CEA. During the earthquake emergency, such information service, provided to the government and the public, was far from satisfactory. This situation was partly because the upgraded seismological monitoring system had just begun its testing-mode operation less than 2 months before the Wenchuan earthquake occurred. The failure included the rapid loss estimation (e.g. fatalities) with a highly underestimated result due to the use of a 'point source' model; the mis-triggering of an event report of 'a $M_L3.9$ earthquake in Beijing' by the seismic waves of the Wenchuan mainshock recorded at the Beijing Capital-Circle Region seismic network; and the automatic regional CMT solution of the mainshock provided by the CENC which was shown to be incorrect.

As early as in 2007, *The Strategic Plan of National Earthquake Science and Technology Development* (2007–2020),[6] which outlines the scientific challenges and

[6] http://www.cea.gov.cn/manage/html/8a8587881632fa5c0116674a018300cf/_content/10_01/28/1264640870755.html

suggestive actions in science and technology for the reduction of earthquake disasters, put seismological information service as one of the seven R. and D. priorities.[7] CEA has been providing the report of earthquakes to the government and the public for a long time. The above mentioned failures of such reports during the emergency response stage of the Wenchuan earthquake were mainly because of dual reasons: On one hand, in the past time, the seismological information service was only treated as a part of the scientific research or the routine observation, and in lack of an overall design in the perspective of systems engineering; On the other hand, entering the time of information, the public has an increasing need of the timeliness, quality, and contents of the information provided by seismological agency. After the Wenchuan earthquake, taking these lessons into account, CEA initiated tremendous efforts to improve the seismological information service. The improvement has been in progress since then based on the up-to-date science and technology. Experiences of the international advanced seismological services, such as the National Earthquake Information Center (NEIC) of the US Geological Survey (USGS), were also investigated. This endeavor was tested as successful during the emergency response to the 2010 Yushu $M_S7.1$ earthquake, and was further tested by the Lushan earthquake.

6.3.2 The Infrastructure and Organization Bases for the Seismological Information Service

The seismological products are, to a large extent, based on the seismological monitoring system in operation. Since the turn of the centuries, infrastructures for seismological observation and monitoring have been developed and modernized considerably in the China mainland.[8] The 'China Digital Seismological Observation System' was completed and began operation at the end of 2000, which includes the National Digital Seismograph Network, 20 regional digital seismograph networks, and a mobile digital seismograph network consist of 100 portable digital seismographs. Between 1999 and 2001, the Beijing Capital-Circle Digital Seismograph Network (covering Beijing Municipality, Tianjin Municipality, and Hebei Province) with real-time data transmission was established, consisting of 107 seismic stations. The 'China Digital Earthquake Observation Networks Project', consisting of 6 components (3 networks and 3 systems): a geophysical/geochemical-anomaly monitoring network, a digital seismograph network, and a digital strong motion network; an

[7] As a reference, priorities identified for the reduction of earthquake disasters include: (1) Theory and techniques for earthquake monitoring; (2) Continental tectonics and 'continental' earthquakes; (3) Oceanic seismology; (4) Earthquake forecast and prediction; (5) Preparedness of earthquake disasters; (6) Techniques for earthquake emergency response and rescue actions; and (7) Seismological information service to scientific communities and the public.

[8] Chen JM Earthquake disaster reduction in developing China, Keynote Presentation, the 14th World Conference on Earthquake Engineering (WCEE), October 12–17, 2008, Beijing, China, electronic version available at: http://www.iitk.ac.in/nicee/wcee/article/14_K002.pdf

active fault mapping system, an earthquake emergency commanding system, and an earthquake information service system, was launched in June 2004, and passed the acceptance inspection in April 2008, just one month before the Wenchuan earthquake.

CEA distributed the jobs of the production of the seismological information products among its institutions, which has established a mechanism to work in a 7 × 24 mode on routine basis since 2009. Chinese version of the 'earthquake poster' has been produced and published online since then (see Fig. 6.2 as an example), which is similar to the earthquake summary poster of the USGS (Fig. 6.3). Similar to what has been down in other agencies (for example, WAPMERR, see Fig. 6.4), quick estimation of the losses is also provided based on seismological monitoring.

6.4 Discussion: From Basic Science to Application—The Technology Readiness Level Related to the Seismological Information Service

CEA provides different kinds of seismological information service to the government and the public (Wu et al. 2013), although not all these information services are expressed by 'scientific products'. Not only CEA but also other agencies such as the Ministry of Civil Affairs have established the system for serving the disaster emergency response (Li et al. 2013). For the maintenance and quality control of such information services, one needs not only the concepts and tools of seismology but also the concepts and tools of systems engineering. That is, the seismological monitoring system has to be treated as a 'machine' of information service, which is not the simple summing up of all the works by individual scientists and institutions.

Treating a seismological monitoring system as a 'machine' of information service is by no means a new idea. It is well-known that the International Monitoring System (IMS) of the CTBTO PrepComm has long been using the concepts and tools such as System Performance Evaluation (SPE) and Key Performance Indicators (KPI) in its maintenance and management. If a seismological agency could be regarded as a 'machine' producing seismological information to the government and the public, then the contents of the information, its timeliness and quality, have to be designed and managed carefully according to the scientific and technological capacity and the needs of the public. How the results of seismological studies could be transformed into the information which is relevant to the public in the reduction of earthquake disasters, and what are the optimized ways for disseminating such relevant information on routine basis are important issues to be considered, for which the concept of technology readiness level (TRL) may be of help.

TRL[9] is a measure used to assess the maturity of evolving technologies, including devices, materials, components, software, and work processes, during its development. When a new technology is first conceptualized, generally it is not suitable for immediate practical application. Instead, new technologies are usually subject to

[9] See: http://en.wikipedia.org/wiki/Technology_readiness_level

(a)

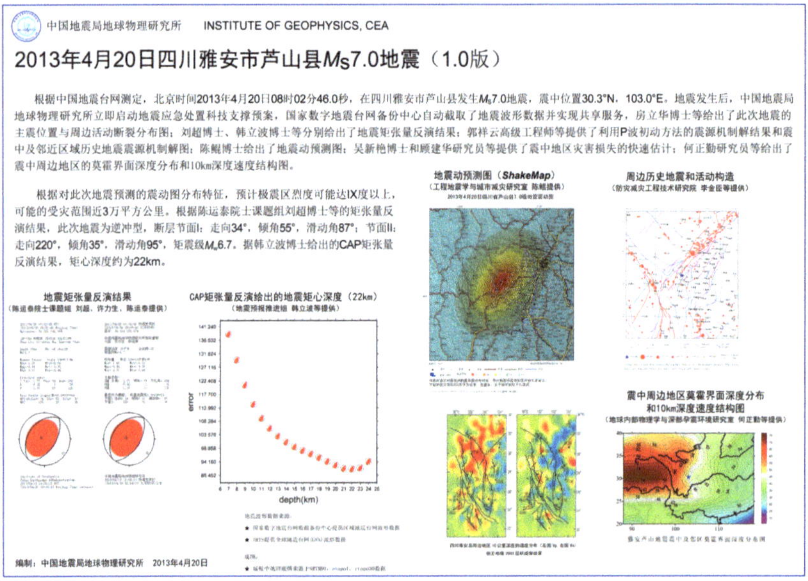

(b)

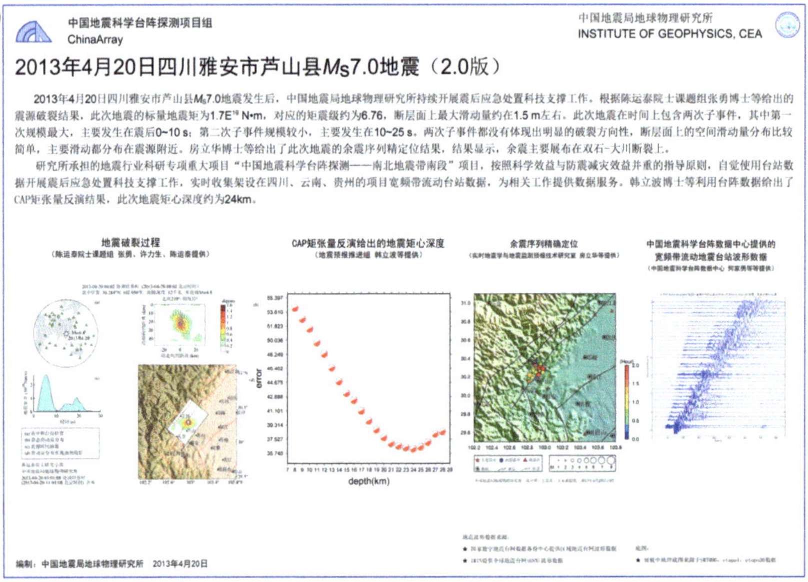

Fig. 6.2 Earthquake posters of the Lushan earthquake: an example of the earthquake posters online, produced by the Institute of Geophysics, CEA (See: http://www.cea-igp.ac.cn/tpxw/index.shtml). *Top* Version 1.0 earthquake poster (http://www.cea-igp.ac.cn/tpxw/266824.shtml) which includes (from *left* to *right*) the focal mechanism, focal depth, ShakeMap, historical earthquakes with seismo-tectonic background, and structure of the seismic source region. *Bottom* Version 2.0 earthquake poster (http://www.cea-igp.ac.cn/tpxw/266810.shtml) which includes (from *left* to *right*) the rupture process, focal depth, aftershock sequence, and seismic recordings of the mobile seismic array (the ChinArray)

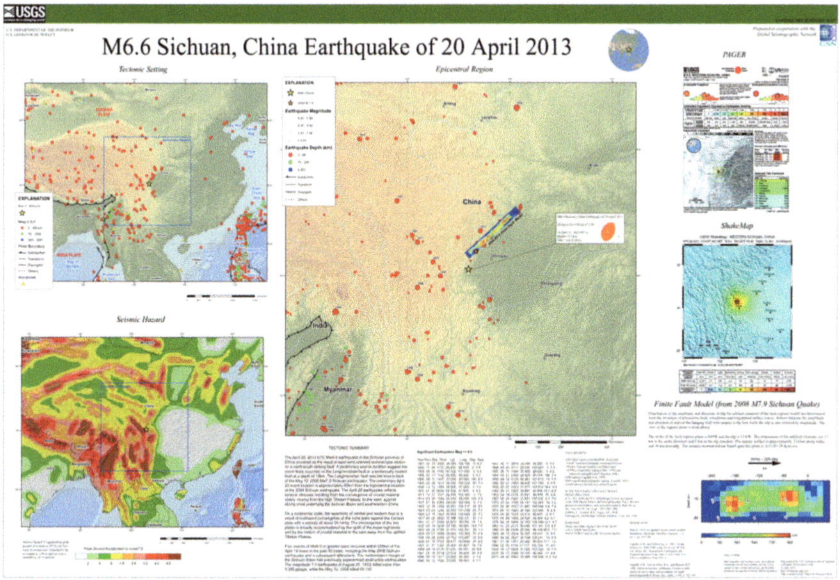

Fig. 6.3 Earthquake summary poster of the NEIC of USGS (http://earthquake.usgs.gov/earthquakes/eqarchives/poster/2013/20130420.jpg) for comparison, showing the distribution of historical earthquakes and zonation of seismic intensity (*left*); relation with the 2008 Wenchuan earthquake (*middle*); and the information of the Lushan earthquake (*right*, with the intensity distribution on the *top*, and ShakeMap in the *middle*. Slip distribution of the Wenchuan earthquake is provided on the *bottom*)

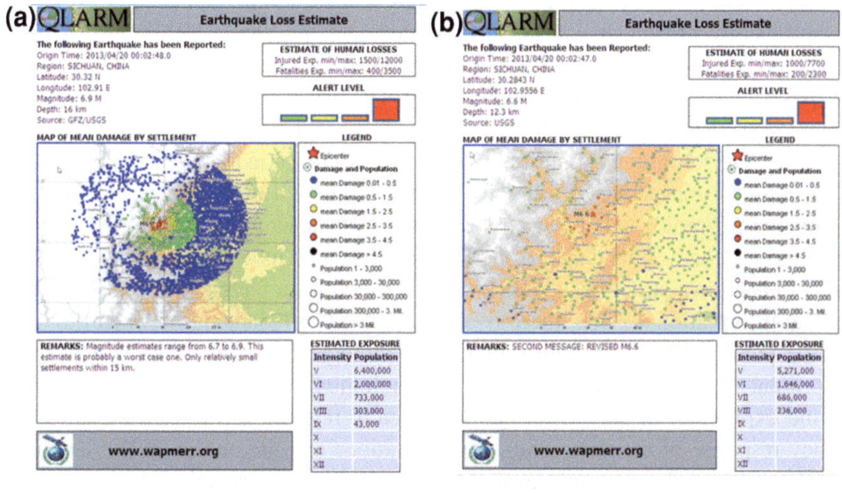

Fig. 6.4 Earthquake loss estimate by WAPMERR. The first estimate **a** gives the expected fatality 400–3500, and the second estimate **b** gives 200–2300. Actual fatality hits the lower bound of the second estimate. The expected fatality estimated by IGPCEA is 100–500 on the first day after the Lushan earthquake

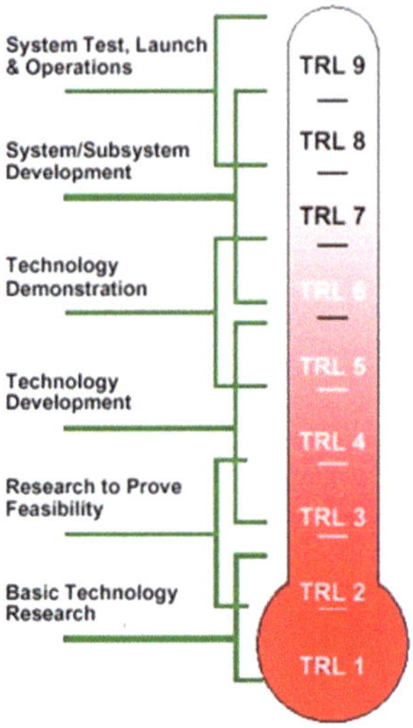

Fig. 6.5 The TRL defined by NASA (See: http://en.wikipedia.org/wiki/Technology_readiness_level). One of the analogues in seismological information service is the centroid moment tensor (CMT) inversion, in which TRL-1 to 2 indicates the formulation of the concept of CMT; TRL-3 to 4 indicates the realization of the inversion algorithm, especially the Green's function database which needs not only the algorithm but also Earth structure information; TRL-5 to 6 indicates the testing-mode inversion procedure towards the information service on routine basis; TRL-7 to 8 indicates the testing for a specific region of the reliability and timeliness of the result; and TRL-9, plus the guidance for interpreting the CMT messages, indicates the validation of the system

experimentation, refinement, and increasingly realistic testing. Once the technology is sufficiently proven, it can be incorporated into a system. TRL was originally developed by NASA in the 1980s and was adopted by different agencies for different purposes. In some countries, such like China, TRL has entered the system of national technique standards.[10] Figure 6.5 shows the TRL standard of NASA, with the correspondence to seismological information service discussed in the figure captions.

It may be unnecessary to directly use the methods for evaluating the TRLs to a specific seismological approach, but to make the seismological information service more effective, the concept of TRL, together with its related concepts, such as work breakdown structures (WBS), may be used as a reference. Indeed in the scientific information service, there is always a need of balancing between the need of the public and the capability of the seismological agency. What is especially important is the idea which piece of seismological information, and which way of the expression of such information, is really relevant to the earthquake emergency response of the emergency administrations and the public. In the language of TRL, not all the presently available scientific results are suitable to act as the emergency-service products.

[10] General Rules of Science and Technology Research Projects Evaluation (GB/T 22900-2009).

6.5 Summary

In this chapter we analyzed the scientific products serving the earthquake emergency response, including the fast reports of the location, magnitude, focal mechanism, and rupture process of the mainshock, refined location of the aftershock sequence, ground motion characterization and loss estimation, and background tectonic information, among others. Such products were produced based on the infrastructure for seismological monitoring, and the organization of the task forces within the CEA institutions. We introduced the development, the status quo, and the on-going works related to the emergency-oriented seismological information service. We used the language of the technological readiness level (TRL) to describe the relation between basic researches and the information service.

Nowadays the improvement of the scientific products as well as the seismological information service to the public is still underway. Aiding the earthquake information service not only for earthquake emergency but also for earthquake disaster preparedness, started from 2010, CEA launched the Social Earthquake Safety Service Engineering, based on the nodes of regional and local earthquake administrations as well as the local governments. Apparently there is still a large room for improvement, if the seismological information service is viewed in a systems engineering and social perspective, especially in our digital or information time.

References

Fang LH, Wu JP, Wang WL, Lü ZY, Wang CZ, Yang T, Cai Y (2013) Relocation of mainshock and aftershock sequences of M_S7.0 Sichuan Lushan earthquake. Chin Sci Bull 58:3451–3459. doi:10.1007/s11434-013-6000-2

Li ZF, Wu LX, Wang R, Wang QL, Zheng S, Xu ZH, Li J (2013) Rapid assessment of earthquake disaster and rethink of M_S7.0 Lushan earthquake. Sci Technol Rev 31(12):31–36. doi: 10.3981/j.issn.1000-7857.2013.12.005 (in Chinese with English abstract)

Liu RF, Chen YT, Zou LY, Chen HF, Liang JH, Zhang LW, Han XJ, Ren X, Sun L (2013) Determination of parameters for the 20 April 2013 Lushan M_W6.7(M_S7.0) earthquake in Sichuan Province. Acta Seismol Sin 35:652–660. doi:10.3969/j.issn.0253-3782.2013.05.004 (in Chinese with English abstract)

Wu ZL, Ma TF, Jiang H, Jiang CS (2013) Multi-scale seismic hazard and risk in the China mainland with implication for the preparedness, mitigation, and management of earthquake disasters: an overview. Int J Disaster Risk Reduction 4:21–33

Xu LS, Yan C, Zhang X, Li CL (2013) Where did the Lushan M_S7.0 earthquake occur in the world? Chin J Geophys 56:2982–2993. doi:10.6038/cjg20130912 (in Chinese with English abstract)

Zhang Y, Xu LS, Chen YT (2013) Rupture of the Lushan 4.20 earthquake and preliminary analysis on the disaster-causing mechanism. Chin J Geophys 56:1408–1411. doi:10.6038/cjg20130435 (in Chinese with English abstract)

Zhao CP, Zhou LQ, Chen ZL (2013) Source rupture process of Lushan M_S7.0 earthquake, Sichuan, China and its tectonic implication. Chin Sci Bull 58:3444–3450. doi:10.1007/s11434-013-6017-6

Chapter 7
An Earthquake Occurred in the Time of the Internet and the Microblog/Twitter

Abstract The Lushan earthquake occurred in the time of the Internet and the Microblog/Twitter, providing some fresh phenomenology for understanding the impact of earthquakes on our society.

Keywords Internet · Microblog or Twitter · Citizen seismology · Statistical physical models of emergency evacuation

7.1 The Microblog (*Weibo*)

7.1.1 Introduction

In recent years, the Internet and Twitter have been used to detect felt earthquakes and map the distribution of intensity (Earle et al. 2010, 2012; Crooks et al. 2013).[1] The application of the Internet and Twitter in earthquake detection not only assists the collecting of the information of earthquakes but also enhances the public participation in the reduction of earthquake disasters, forming an active branch of 'citizen seismology' (Bossu et al. 2008).[2]

Sina Weibo, one of the Chinese versions of Twitter, is one of the most popular microblogging websites in the mainland of China, which is used by over 1/3 of the Chinese Internet users. By the time of March 2013, *Sina Weibo* users have reached 536 million, becoming China's largest social media platform. This platform plays an active role in the reporting of the details of the Lushan earthquake.

This chapter is written by Ma TF and Wu ZL (Institute of Geophysics, CEA), with the assistance of Jiang H.

[1] Micro-blog (Weibo) is a Chinese version of Twitter. Weibo is the Chinese word in which 'Wei' = micro and 'Bo' = blog.

[2] See: http://www.citizenseismology.eu/

DOI: 10.1007/978-981-4585-15-6_7,

China Earthquake Networks Center (CENC) has launched its *Seismological Weibo* since 2012 (Hou 2013). This official *Dizhen* (earthquake) *Weibo* (microblog) ID was registered on May 28, 2012. Messages of automatically determined earthquake parameters can be distributed via this ID, together with the epicenter map based on *Google Maps*© or *Baidu Maps*©. Up to March 2013, a mechanism of routine message distribution has been established, with about 8 messages per day in a round-the-clock mode. Amongst all the messages, rapid earthquake reports account for about 55 %, knowledge of earthquake science about 15 %, interaction with the fans about 10 %, and others about 20 %. The *Seismological Weibo* has owned more than 1.4 million fans, ranking the 70th among the 65,000 official *Weibo* IDs in the mainland of China. The *Seismological Weibo* also won several awards in 2012 (from Tencent.com, Sina.com, and People.cn) due to its public influence and technical innovation. The Lushan earthquake is the first strong earthquake testing the performance of the *Seismological Weibo,* as well as the Weibo users' network.

7.1.2 The Performance of the Microblogs in Reporting the Lushan Earthquake

The Lushan earthquake occurred at 08:02 a.m. local time, on Saturday April 20, 2013. One minute after the earthquake, the *Seismological Weibo* published the automatic detection of the event, with location information 'near Ya'an City, Sichuan', 30.1 °N and 103.0 °E, and the quick magnitude determination result *M*5.9. At 08:14 a.m., the location was modified as 'Lushan County, Ya'an City', with preliminary focal depth 13 km, and revised magnitude *M*7.0.

At 08:29 a.m., the China International Search and Rescue Team (CISAR) appealed people in the earthquake-stricken area to report their location and the damage via Microblogs with cell phones if possible. That post had been forwarded nearly 43 million times by 10:00 p.m. on Saturday. On April 20, there were nearly 66 million items related to the Lushan earthquake appeared on *Sina Weibo*. The most re-tweeted *Weibo* messages contain three main topics: rescue information, earthquake relief knowledge, and donation.

For governmental agencies, Microblog users are also a good feedback platform. On April 20, through Microblogs it was complained that the Ya'an highway still took charges, resulting in traffic jams after the earthquake. In response to this message, authorities quickly corrected the mistake by stopping charging.

In addition, through the microblog Location Based Services (LBS) positioning, the authorities can get detailed information about the stricken areas, which became a milestone in China for the role of social media in earthquake and other disasters.

Figure 7.1 shows the *Sina Weibo* items related to the Lushan earthquake during the first 24 h after the occurrence of the earthquake. Figure 7.2 shows the number of reports in *Sina News* related to the Lushan earthquake during the first 24 h after the mainshock. The temporal dependence is, to much extent, similar to the Internet jam and understandable in the perspective of the dynamics of networks (Bossu et al. 2008).

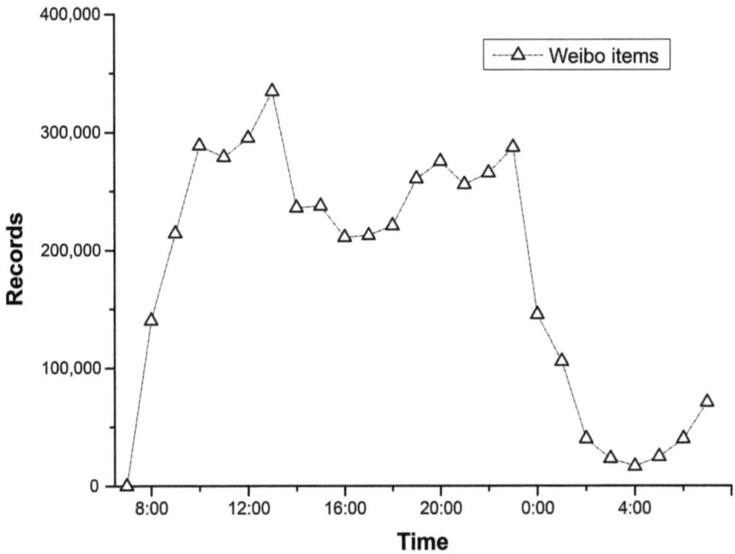

Fig. 7.1 *Sina Weibo* items related to the Lushan Earthquake during the first 24 h after the occurrence of the earthquake (*Source* http://www.weibo.com.cn/). Note that in the mainland of China, the daily CCTV *Xinwen* (News) *Lianbo* (Joint program of the news TV stations of the whole mainland of China) is firstly broadcast at 19:00 p.m. on routine basis, reflecting the official response to the news within the day. Whether the second peak in the curve is related to this news program is subject to future investigation

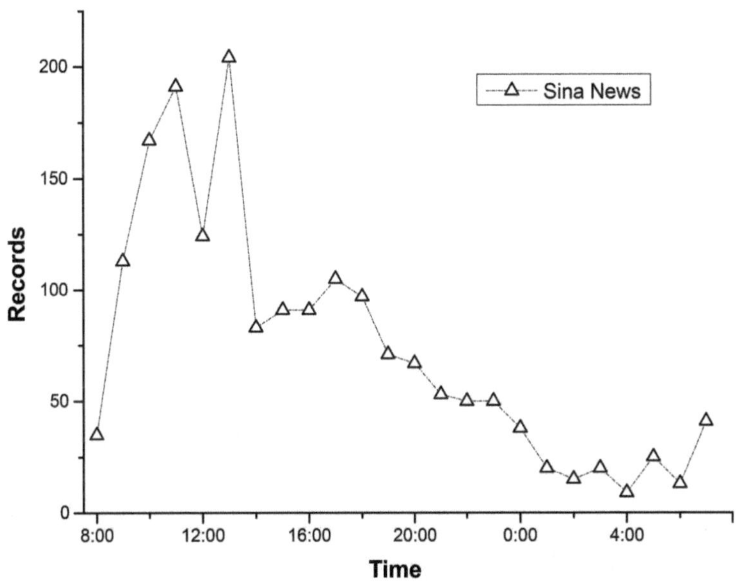

Fig. 7.2 Number of reports in *Sina News* related to the Lushan earthquake during the first 24 h after the occurrence of the earthquake (*Source* http://news.sina.com.cn/z/yadzh2013/)

But how to understand the difference in the characterized time scales, as shown in Figs. 7.1 and 7.2, respectively, is one of the open questions to investigate, referring to the captions of Fig. 7.1.

7.1.3 Discussion

When the M_S8.0 earthquake struck Wenchuan and Beichuan in 2008, microblogs had not yet become popular, and played a minor role in the rescue and relief actions. In contrast, in the emergency response to the Lushan earthquake, microblogs served as a key platform for people to obtain and disseminate the timing information.

The Lushan earthquake emergency also demonstrated the limitation of the social-networking media, namely rumors and/or exaggerated/biased information. For example, a series of photos of collapsed buildings and injuries appeared on *Sina Weibo* in Saturday morning were revealed to be the 'out news' of the 2008 Wenchuan earthquake. On the other hand, however, the revealing of such mis-information also demonstrates the potential of the social-networking media to detect and correct the mistakes. The problem is how to avoid the mis-guidance of such mis-information in the emergency response stage, and how to foster the capability of the self-correction of the 'Microblog society'.

The *Seismological Weibo* informed its fans of the knowledge of earthquake parameter determination, especially the relation between the quick-and-dirty parameters and the well-determined parameters.[3] Note that it was the 2008 Wenchuan earthquake that marked the first time for Chinese seismological agency to publish the revised parameters of an earthquake, the Lushan earthquake marked a new change of public understanding of seismological monitoring—although the change was apparently not as timing as it should be comparing to other countries/regions.

7.2 Death Toll Message from the Internet

7.2.1 Introduction

Sina News launched the *Lushan earthquake special* on April 20, 2013, in which some of the reports were related to the deaths found by the rescue teams. Temporal variations of fatalities reported after earthquakes and/or earthquake-generated tsunamis have been studied for years. In 2001, investigating the 1999 Chi-Chi earthquake, Tsai et al. (2001) analyzed the time-dependent cumulative numbers of killed and missing people found by search and rescue teams. Gao and Jia (2005) tried to model the temporal changes of found deaths by earthquake rescue

[3] http://www.best-news.us/

operations using a polynomial function with degrees-of-freedom 5. Modifying the data-fitting function of Gao and Jia (2005), and based on the cases of the 2004 Indian Ocean tsunami and the 2005 Pakistan earthquake, Liu and Wu (2005) proposed a simple model with degrees-of-freedom 2. More cases were collected later, obtaining that the rescue-efficiency parameter in the model of Liu and Wu (2005) is magnitude-dependent (Zhao et al. 2008). The model was also used for the early estimate of the final death toll (Li et al. 2011). The Lushan earthquake provides another case for testing the model.

7.2.2 Death Toll Message and Its Time Dependence

In the model of Liu and Wu (2005), the time-dependent reported number of deaths at time t, $N(t)$, can be represented in a simple form by

$$N(t) = N_0[1-\exp(-\alpha t)] \tag{7.1}$$

in which N_0 and α are constants, which is shown to be a good approximation of the real cases for 12 earthquakes and/or earthquake-generated tsunami (Zhao et al. 2008). Equation (7.1) can be simply understood by taking the time derivative of both sides:

$$\frac{dN}{dt} = \alpha(N_0-N) \tag{7.2}$$

in which term dN/dt on the left-hand side is the rate of discovering dead bodies, and term $(N_0- N)$ on the right-hand side is the dead bodies remained undiscovered, where constant N_0 is the final death toll. Proportional relation between these two quantities, dN/dt and $(N_0 - N)$, is straightforwardly understandable in rescue operation: Qualitatively, the more the bodies of deaths remained to be discovered, the higher the probability for the rescue team to discover them; The coefficient α linking these two terms reflects the efficiency of the rescue practice, depending on the site condition and the rescue facilities. But more importantly Zhao et al. (2008) and Li et al. (2011) observed that the orders of magnitudes (that is, the logarithm) of coefficient α decreases linearly with the orders of magnitudes of the radiated energy of the earthquake (as represented by its surface wave magnitude M_S), as shown in Fig. 7.3. Although with significant uncertainties, this relation reflects the dependence of the difficulties the rescue action encounters on the size of the earthquake.

7.2.3 Result and Discussion

Figure 7.4 shows the deaths found, reported by *Sina News* at different time, and the fitting curve using Eq. (7.1). From the figure it may be seen that Eq. (7.1) well describes the real situation. The fitting gets N_0 being about 200, which is in

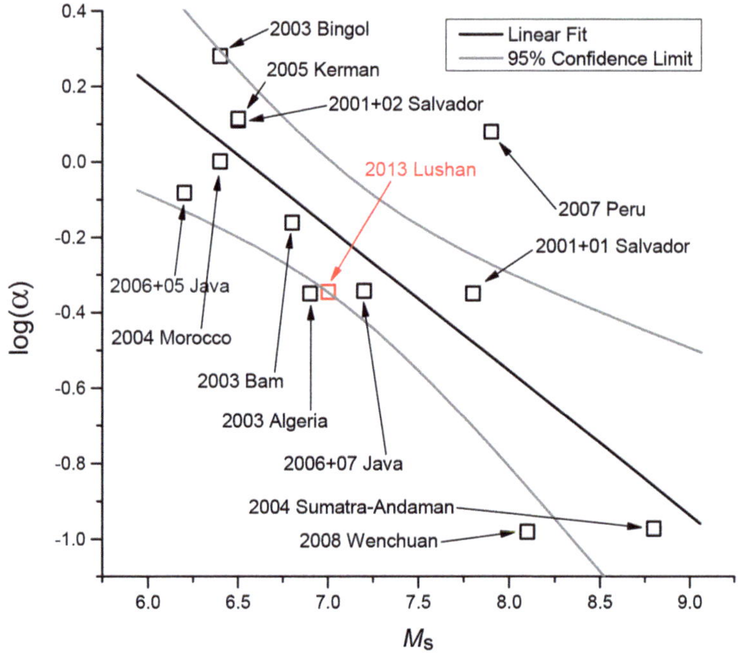

Fig. 7.3 Relation between coefficient α and magnitude M_S (Modified from Li et al. 2011)

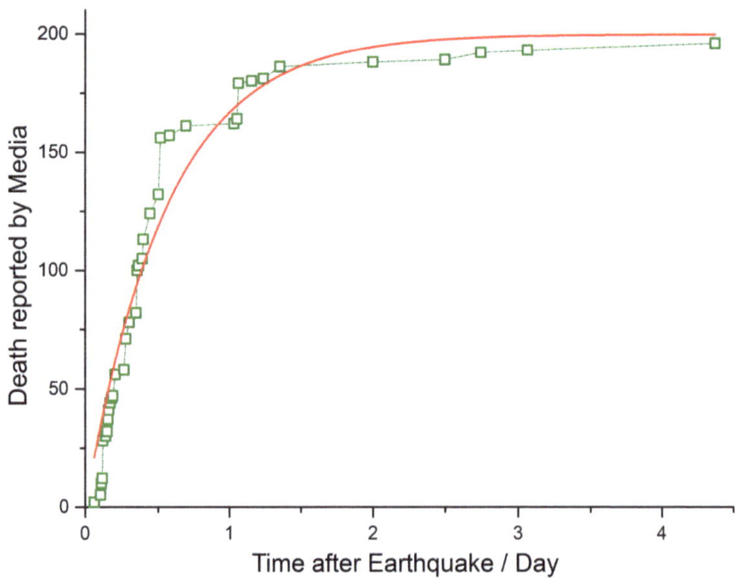

Fig. 7.4 Deaths found, reported by *Sina News* at different time (*squares*) and its fitting curve (in *red*) using Eq. (7.1) (*Source* http://news.sina.com.cn/z/yadzh2013/)

consistence with the real case. The rescue efficiency coefficient $\alpha \approx 0.45$. For the convenience of comparison in Fig. 7.3, $\text{Log}_{10}(\alpha) \approx -0.35$. From Fig. 7.3 it can be seen that the point at which $\text{Log}_{10}(\alpha) \approx -0.35$ and M_S being 7.0 falls into the 95 % confidence limit of the fitting. On the other hand, if M_S is taken less than 7.0, say, 6.6, then the point deviates from the fitting line. This provides an indirect evidence (of course with clear cautions in mind of its limitation) that M_S being 7.0 (although with a smaller M_W) is due to seismological reasons rather than the problems of the measurement (See Chap. 1 in this Brief).

7.3 Classroom Evacuation Recordings

7.3.1 Introduction

After the earthquake, several video recordings were broadcast through TV and the Internet (Ma et al. 2013). These records provide vivid materials for the public understanding of earthquake emergency. The real-life disaster records are also important observational data for constraining the statistical physical models of emergency evacuation.

In recent years, quite a few self-organization phenomena in pedestrian streams have been successfully reproduced by statistical physical models such as molecular dynamics models, lattice gas models, or cellular automaton. For review of the related models, see Zheng et al. (2009). Theoretical results of these models are of essential importance not only for the design and implementation of transportation and building environment but also for the emergency evacuation during natural and/or man-made disasters such as earthquakes and fires. Important is that, if the model/s were wrong, then the result of the design would be disastrous (see, e.g., Soomaroo and Murray 2012).

In constraining these models by empirical analysis of observational or experimental data, video tracking has shown to be a valuable tool (Liu et al. 2009; Helbing et al. 2000b) On the other hand, however, for the escape panic in real-life disasters, except a few cases (e.g., Johansson et al. 2008), such observational records are still rare. This is firstly due to technical difficulties because disasters such as earthquakes are difficult to forecast. At the mean time, ethical reason is another obstacle in obtaining such recordings because near-real-life disaster experiments, either with animals or with human being, are essentially problematic due to the danger of potential injury.

Considering the importance of real-life disaster records in the physical modeling of escape panic (Helbing et al. 2000, 2005, 2007), the video recordings of earthquake emergency evacuation are of unique values not only for the public understanding of earthquake disaster reduction but also for statistical physical studies. Preliminary investigation (Yang et al. 2011) showed that the difference between the behavior of pedestrians in real-life disaster panic and that in mimic exercises is significant.

Fig. 7.5 One frame of the video recording of the classroom evacuation in the Mingshan Middle School (*Source* http://v.youku.com/v_show/id_XNTQ2MDQzNDQ0.html)

7.3.2 *Observation*

Figure 7.5 shows one frame of the video recording of the classroom evacuation in the Mingshan Middle School, just in the epicentral region of the Lushan earthquake. CCTV broadcast this video recording in its news program. In the figure is a characterized emergency evacuation scene: an exit in the classroom used for evacuation during the earthquake, and the evacuation was organized by the teacher. Figure 7.6 shows the evacuation time and direction (red arrows) of different students. Note that in the middle row of the desks, students have two optional directions to evacuate, namely leftward or rightward, respectively. The ends of the red arrows divide the two groups of students escaping leftward and rightward, respectively. Such scene is typical in the classroom experiments, albeit it is not mimic disaster situation but the real-life disaster.

In the processing of the video recordings, the *Total Video Converter*© was used to transform the dynamic videos into discrete frames. The measurement was conducted manually. The measurement is rough due to the quality of the video camera and the ground shaking caused by the earthquake. Despite the apparent shortcomings, result obtained through the measurement reveals some interesting features of escape panic during real-life earthquake emergency. Figure 7.7 shows, at the exit, the arrival time against the order of the student arriving, which is an important measure for comparison with theoretical models. Previous study using the video recordings of the 2008 Wenchuan earthquake (Yang et al. 2011) shows that this relation is different from what has been predicted by simulated experiments and theoretical models (Nagai et al. 2005, 2006; Zhang et al. 2008), that mimic experiments and models predict a linear relation, but the real-life relation is not linear. However, for the case shown in Figs. 7.5, 7.6, 7.7 and 7.8 the relation is piece-wise linear, being divided by two linear stages at about 25 s. Reason for this change is that the evacuation was organized by the teacher, and before the earthquake, due to the lessons of the Wenchuan earthquake, students were well

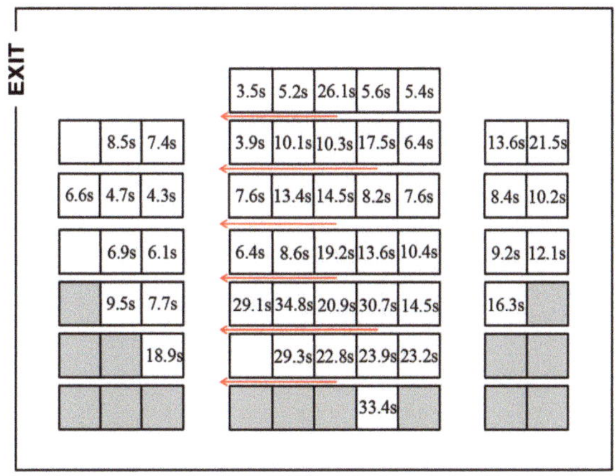

Fig. 7.6 Evacuation time and direction (*red arrows*) of different students. The *gray blocks* indicate blind spots for the video monitor, the *blank areas* means the recordings are difficult to identify. Exit is to the *top left*

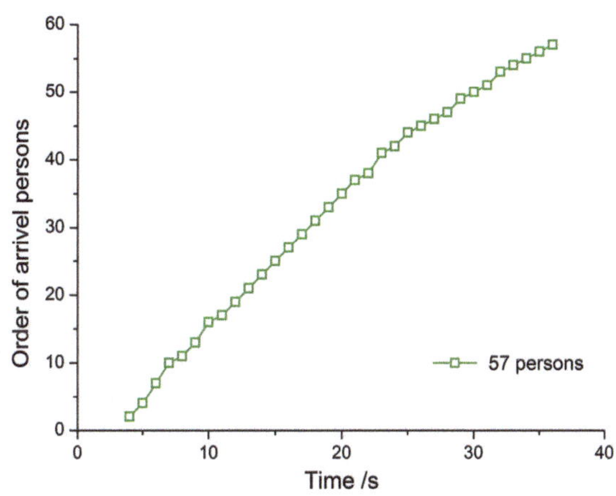

Fig. 7.7 The arrival time against the order of the student arriving: the Mingshan Middle School

trained for earthquake evacuation. It turns out that trained and untrained crowds may have different behavior during the panic, reflecting the advancements after the Wenchuan earthquake in public understandings.

At the mean time, it should be noticed that it is not the case that students in all the schools were well trained, which is a large space for improvement. Figure 7.8 shows, as a comparison, the situation of another middle school, further from the epicenter comparing to the middle school in the former example. Similar to what

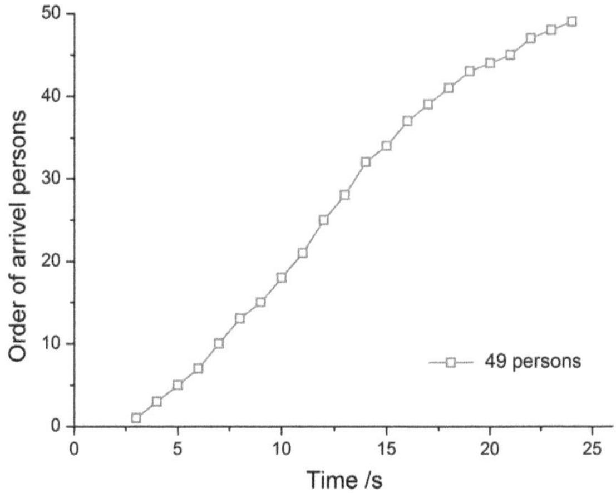

Fig. 7.8 The arrival time against the order of the student arriving: the Jinchuan Middle School (*Source* of evacuation video http://v.ku6.com/show/Yp-cFyFXyd790zZH3dO6xA... html)

happened during the Wenchuan earthquake, the relation between the arrival time against the order of the student arriving deviates from a simple linear relation.

7.3.3 Discussion

In recent years, with the development of economy and society, there have been more and more civilian monitoring videos deployed in different places, occasionally recording the scenes of earthquakes. These real-life disaster records are potentially useful for studying earthquake intensity, strong ground motion, and the physics of escape panic (Hori et al. 2000; Sutoh et al. 2003; Johansson et al. 2008). In the studies aiming at the design of transportation and buildings as well as fire disaster management, databases of video records have been developed and used within a wide range (Fahy and Proulx 2001a, b; Galea et al. 2005). Video records of survivors are also used for studying tsunami (Fritz et al. 2006). Digest of this new type of data needs more works in future. And calling for the collection and on-linc sharing of such data may bc of practical importancc both for public participation and for scientific studies.

7.4 Summary

In this chapter we analyzed a new type of data recording the phenomenology of earthquakes. The microblog messages, the Internet reported death toll, and the video recordings of emergency evacuation, although still having many shortcomings as a 'scientific' database, are brand new to earthquake studies and are associated with our time of the Internet and Microblogs/Twitter. One of the common characteristics of these data is that none of them are purposely deployed for earthquake emergency, and all of them are maintained mainly by the public. These civilian recordings provide interesting data for the study of earthquake-related scientific problems, and have a noticeable potential to be studied carefully due to its public nature which provides a tremendous number and range of 'sensors', with some new phenomena hidden in this 'ocean of sensors' to be discovered in near future.

References

Bossu R, Douet V, Rives S, Marin S, Aupetit M (2008) Internet users as seismic sensors for improved earthquake response. EOS Trans AGU 89:225–232

Crooks AA, Stefanidis CA, Radzikowski J (2013) Earthquake: Twitter as a distributed sensor system. Trans GIS 17:124–147

Earle PS, Bowden DC, Guy M (2012) Twitter earthquake detection: earthquake monitoring in a social world. Ann Geophys 54:708–715. doi:10.4401/ag-5364

Earle P, Guy M, Buckmaster R, Ostrum C, Horvath S, Vaughan A (2010) OMG earthquake! can twitter improve earthquake response? Seism Res Lett 81:246–251

Fahy RF, Proulx G (2001a) Toward creating a database on delay times to start evacuation and walking speeds for use in evacuation modeling. In: 2nd international symposium on human behavior in fire, Boston, MA, U.S.A, Mar 2001, 175–183

Fahy RF, Proulx G (2001b) Pre-movement time database. http://www.fireevacuation.ru/pmt/

Fritz HM, Borrero JC, Synolakis CE, Yoo J (2006) 2004 Indian Ocean tsunami flow velocity measurements from survivor videos. Geophys Res Lett 33:L24605. doi:10.1029/2006GL02678

Galea E, Shield J, Canter D (2005) High-rise evacuation evaluation database. http://fseg2.gre.ac.uk/HEED/request/index.html

Gao JG, Jia Y (2005) A study on the time of promulgating earthquake disaster—an index of earthquake rescue ability. Journal of Catastrophology 20:31–35 (in Chinese with English abstract)

Helbing D, Farkas I, Vicsek T (2000a) Simulating dynamical features of escape panic. Nature 407:487–490. doi:10.1038/35035023

Helbing D, Farkas I, Vicsek T (2000b) Panic: a quantitative analysis. http://angel.elte.hu/panic/

Helbing D, Buzna L, Johansson A, Werner T (2005) Self-organized pedestrian crowd dynamics: experiments, simulations, and design solutions. Trans Sci 39:1–24

Helbing D, Johansson A, Al-Abideen HZ (2007) Dynamics of crowd disasters: an empirical study. Phys Rev E 75:046109. doi:10.1103/PhysRevE.75.046109

Hori M, Sutoh A, Saitoh Y (2000) Strong motion measurement using security video cameras. J Struct Mech Earthq Eng 17:47–56 (in English with Japanese abstract)

Hou JM (2013) SeisMicroblog or Dizhen Weibo: One year experience. IAHS-IAPSO-IASPEI Joint Assembly, Gothenburg, Sweden, 22–26 July 2013. http://iahs-iapso-iaspei2013.com/Abstracts.aspx?252482

Johansson A, Helbing D, Al-Abideen HZ, Al-Bosta S (2008) From crowd dynamics to crowd safety: a video-based analysis. Adv Complex Syst 11:497–527

Li YC, Wu ZL, Zhao YZ (2011) Estimating the number of casualties in earthquakes from early field reports and improving the estimate with time. Nat Hazards 56:699–708

Liu X, Song WG, Zhang J (2009) Extraction and quantitative analysis of microscopic evacuation characteristics based on digital image processing. Phys A 388:2717–2726

Liu Z, Wu ZL (2005) A simple model of reported casualties during earthquakes and earthquake-generated tsunamis. Earthq Res China 21:526–529 (in Chinese with English abstract)

Ma TF, Wu ZL, Yang XL (2013) Video recording of classroom evacuation during the April 20, 2013, Lushan, Sichuan, China, earthquake: preliminary analysis and implication for the statistical physical models of escape panic. Wuli (Physics) 42:341–345 (in Chinese with English abstract)

Nagai R, Fukamachia M, Nagatani T (2005) Experiment and simulation for counterflow of people going on all fours. Phys A 358:516–528

Nagai R, Fukamachi M, Nagatani T (2006) Evacuation of crawlers and walkers from corridor through an exit. Phys A 367:449–460

Soomaroo L, Murray V (2012) Disasters at mass gatherings: lessons from history. PLoS Curr. 4: RRN1301. doi: 10.1371/currents.RRN1301

Sutoh A, Hoshiya M, Hori M, Saito Y (2003) Simplified measurement system and an identification procedure of strong ground motion using security video records. In: 9th international conference on applications of statistics and probability in civil engineering, San Francisco, USA, 1559–1564

Tsai YB, Yu TM, Chao HL, Lee CP (2001) Spatial distribution and age dependence of human-fatality rates from the Chi-Chi, Taiwan, earthquake of 21 September 1999. Bull Seism Soc Amer 91:1298–1309

Yang XL, Wu ZL, Li YC (2011) Difference between real-life escape panic and mimic exercises in simulated situation with implications to the statistical physics models of emergency evacuation: the 2008 Wenchuan earthquake. Phys A 390:2375–2380. doi:10.1016/j.physa.2010.10.019

Zhang J, Song WG, Xu X (2008) Experiment and multi-grid modeling of evacuation from a classroom. Phys A 387:5901–5909

Zhao YZ, Wu ZL, Li YT (2008) Casualty in earthquake and tsunami disasters: Internet-based monitoring and early estimation of the final death toll. In: Proceedings of the 14th world conference on earthquake engineering (WCEE), Beijing, 12–17 Oct 2008. Paper ID 09-01-0044

Zheng X, Zhong T, Liu M (2009) Modeling crowd evacuation of a building based on seven methodological approaches. Build Environ 44:437–445

Chapter 8
Locally Characterized Seismic Destruction Phenomenology

Abstract Field investigation of the engineering damages of the Lushan earthquake revealed some new and to much extent unique damage phenomena. The performance of the engineering facilities as well as the lifeline system also provide an opportunity to evaluate the post-Wenchuan-earthquake reconstruction, which will be interesting not only for the local people but also for other places subject to similar seismic risk.

Keywords Engineering damage · Lifeline system · Post-Wenchuan earthquake reconstruction

8.1 Seismic Intensity and Strong Ground Motion

The strong ground motion of the Lushan mainshock was well-recorded by the China National Strong Motion Observation Network System of China (NSMONS), although the number of stations available was not very good. There were 123 stations with epicentral distances from 19 to 770 km recording the three-component accelerograms. Among them, 3 stations are located in the zone with intensity VIII, 2 stations in intensity VII, and 8 stations in intensity VI.[1] The largest recorded peak ground acceleration (PGA) is 1,005.3 cm/s^2 at Station 51BXD along the EW component (Fig. 8.1). This recording may be strongly affected by the site condition at this station which is located on a steep slope rather than flat free-field (Wen et al. 2013; Dai and Li 2013). Meanwhile there are still other 9 stations with PGA larger than 200 cm/s^2, and two of them with PGA larger than 500 cm/s^2.

This chapter is written by Li XJ, Mei ZH, Xie JJ, Wang YS and Liu AW (Institute of Geophysics, CEA).

---

[1] China Earthquake Administration (2013) Intensity map of the 4•20 Lushan, Sichuan, M_S7.0 earthquake. http://www.cea.gov.cn/ (in Chinese).

Z. Wu et al., *Earthquake Phenomenology from the Field*, SpringerBriefs in Earth Sciences, DOI: 10.1007/978-981-4585-15-6_8,

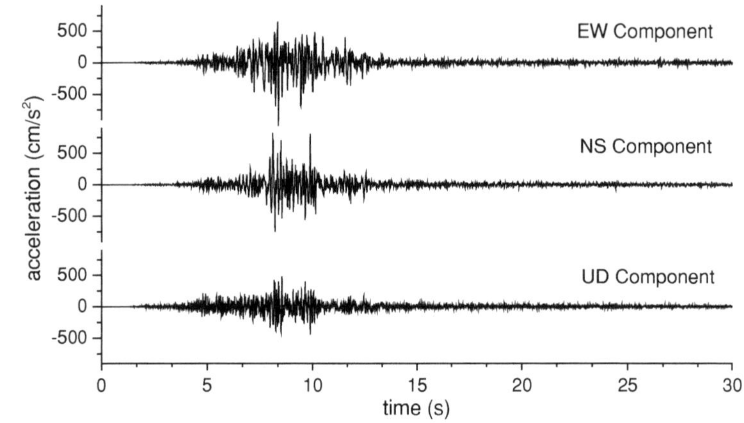

Fig. 8.1 Time histories of acceleration records at station 51BXD (in Baoxing County), which is 19 km NWW away from the epicenter. From the figure it can be seen that peak acceleration appears at 8 s from the first motion

Intensity distribution of the Lushan earthquake is basically based on the field survey of damages. Figure 8.2 shows the typical damages in the zones with different intensity.

Near-source strong ground motions are mostly affected by the faulting mechanism and source effects (Xie et al. 2014). Short-period ground shakings are observed within the vicinity of epicenter, while long-period ground shakings with period $T > 1.0$ s are relatively not so strong, which might help to explain the phenomenon that damages was mainly on the low-story structures (Xu et al. 2013).

8.2 Destruction Phenomenology

The Lushan Earthquake occurred in the west of Sichuan Province, where most of the places were mountainous area which changed to much extent the ground motion and in turn the damage degree of engineerings. In this area, another factor adding the destruction is a kind of cumulative damages, that some of the buildings had already been damaged, not necessarily visible, by the 2008 Wenchuan earthquake, and not yet reinforced because of such 'slight' or 'internal' damages. Fortunately, however, the buildings built after the Wenchuan earthquake performed well during the Lushan earthquake. Figure 8.3 shows an example of such a sharp comparison.

Since the Lushan earthquake, based on field investigations, there have been several publications related to the strong motion, intensity, and engineering damage (Dai et al. 2013; Gao et al. 2013; Gong et al. 2013; Jin et al. 2013; Li et al. 2013; Qu et al. 2013; Sun et al. 2013; Wang et al. 2013a, b, c; Xiong et al. 2013; Xu et al. 2013;

Fig. 8.2 Damages of typical buildings located in different intensity zones. **a** Damage of a frame construction in Baosheng Town, Lushan (Intensity IX). **b** Cracks of a brick masonry building in Feixianguan Town, Lushan (Intensity VIII). **c** Cracks of a brick masonry building in Lieshi Town, Yingjing (Intensity VII). **d** Expanded cracks of a loam wall building with existing cracks in Sanjiao Town, Hanyuan (Intensity VI)

Zhou et al. 2013). In this Brief, rather than comprehensively review these published results, we concentrate our attention on the new and localized experiences related to earthquake disasters.

8.2.1 Residential Buildings

Generally, residential buildings in Lushan and the surrounding area can be categorized into three main structural types, namely wooden structure, brick masonry structure, and steel-reinforced concrete frame structure.

Wooden structure has a high ductile capacity and would perform better than other structures in most cases. Indeed in the epicentral region, most of the wooden structures just had tile-slip damages. However, some wooden structures were heavily damaged, with some of them collapsed, even if near the slightly damaged buildings. Figure 8.4 shows an example of such situation. Investigation showed that most of these heavily damaged wooden structures were infilled with brick walls (Gao et al. 2013). According to Gao et al. (2013), the main causes for this

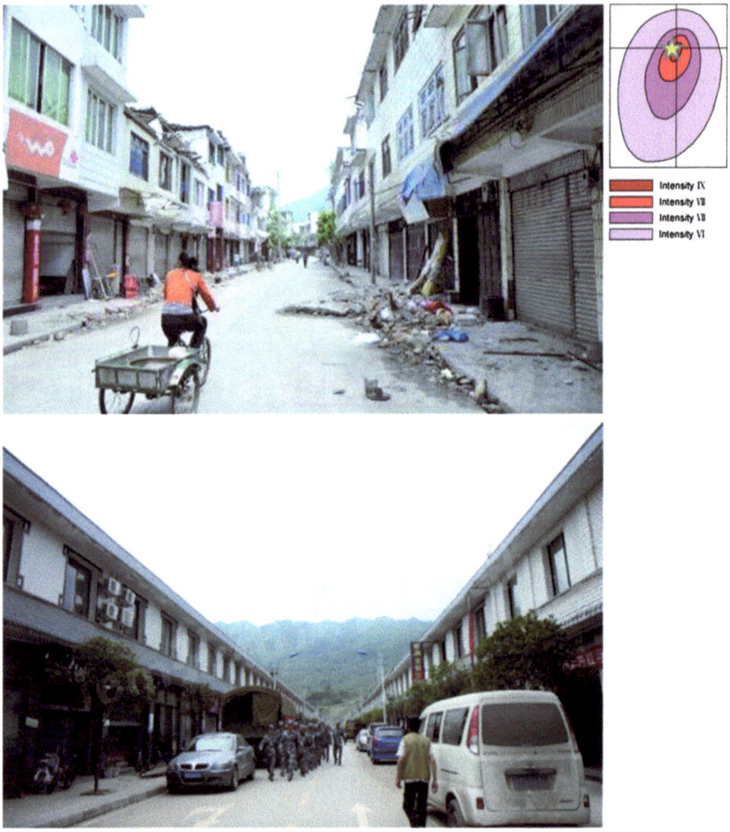

Fig. 8.3 Comparison of the houses built before and after the 2008 Wenchuan earthquake, respectively, at Longmen Town. *Top* Houses built before the 2008 Wenchuan earthquake, in which 80 % of the buildings were heavily damaged, becoming a kind of "standing ruins". *Bottom* Houses built after the 2008 Wenchuan earthquake which performed well during the Lushan earthquake. The indexing map on the *top right* shows the location of pictures (*star*) with the intensity map as background. In the following figures, the indexing map is shown in the same way

kind of 'unusual' damage are: (1) The structures were old and in lack of maintenance, therefore their connections and piers became loose and even corroded; (2) There were brick wall impeded the displacement of the wooden structure, which increased the structure's stiffness and could have caused the stiffness mutation; and (3) Poor designing of the connections between wooden columns and brick walls caused the detachment.

Brick masonry structures are the most severely affected structure types. Due to the locally characterized design of such buildings a new type of destruction appeared, namely the 'standing ruins', indicating the buildings not collapsed but almost lose the capability of supporting (as seen in Fig. 8.5). Wang et al. (2013b, c) and Xu et al. (2013) studied the mechanism of this phenomenon, and found that the fully-restrained brick masonry structure (well designed with ring beam and

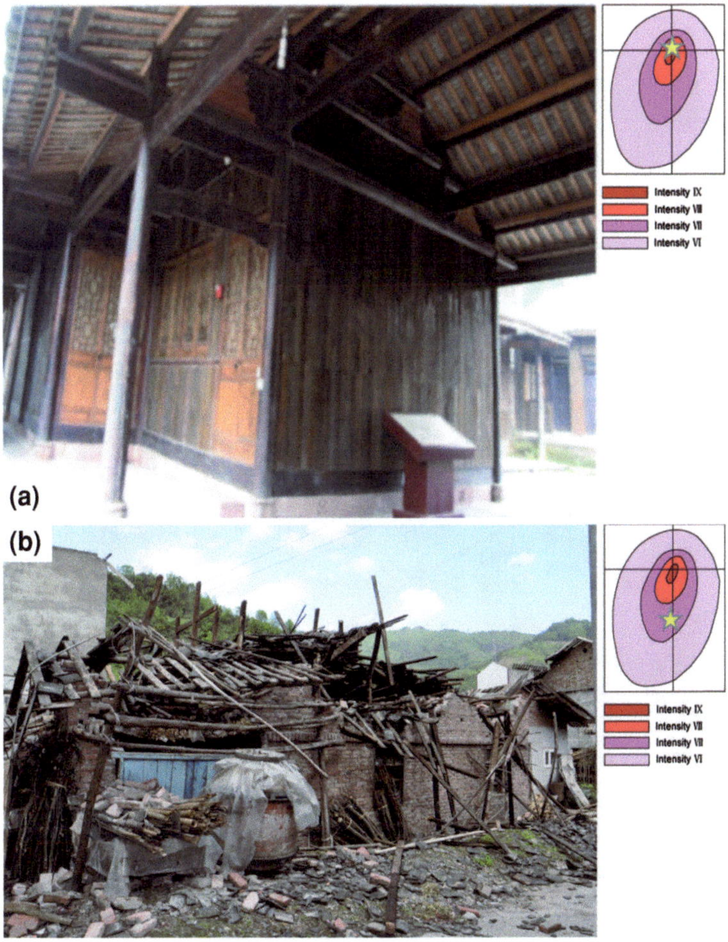

Fig. 8.4 Damage of wooden structures. **a** A traditional wooden structure at Gaohe County which performed well during the earthquake, picture from Gao et al. (2013). **b** A collapsed wooden structure at Rongjing County which was infilled with brick wall

concrete columns) performed much better than un-restrained ones; the pre-cast concrete panel shows its deficiency in resisting earthquake forces; the structure irregularity always increases structure's seismic hazard, which included vertical uneven mass distribution, polygon layout and irregular layout of both vertical and horizontal; and the 'whipping effect' causes additional damage of rooftop projection and adding stories. These lessons are important not only for local reconstruction but also for the places with similar types of buildings.

Steel-reinforced concrete frame structures performed well during this earthquake (Gong et al. 2013). Only frame infilled walls cracked even in the hard-hit areas (Fig. 8.6).

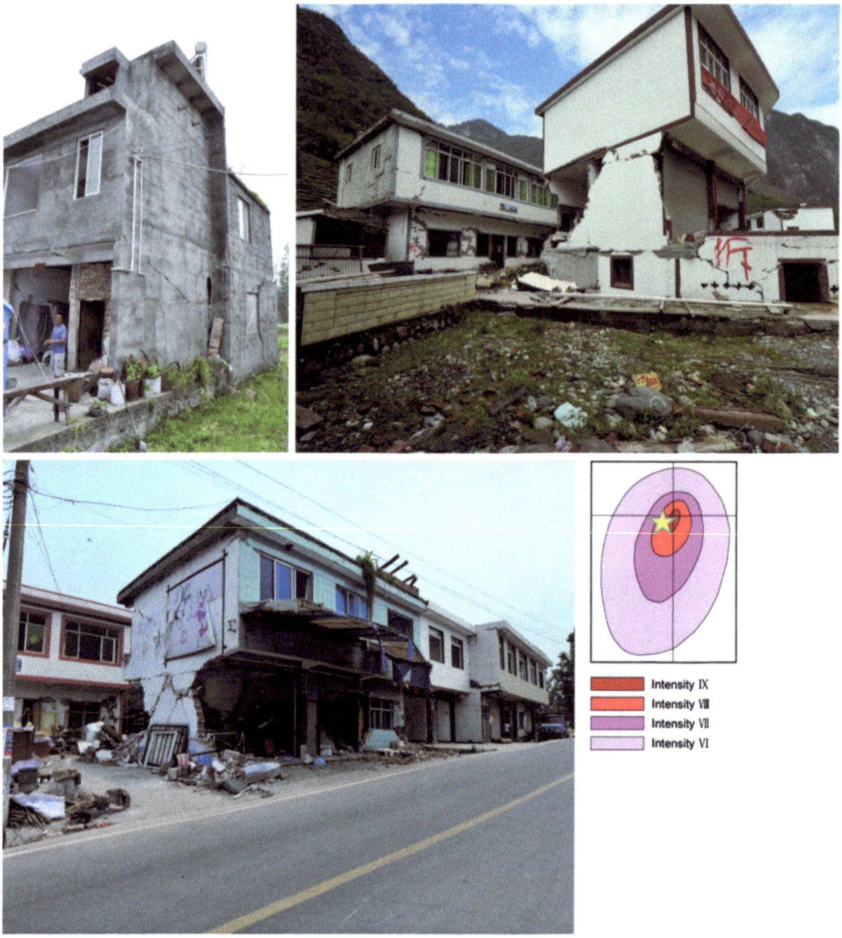

Fig. 8.5 "Standing ruins" located in Longmen Town, Lushan. All these buildings were heavily damaged and irreparable but not collapsed

8.2.2 Public Buildings

Public buildings, which include schools, museums, hospitals, and gymnasia, and so on, always draw people's attention in the earthquake, due to its importance for the emergency evacuation. In Lushan County with intensity VIII, there were some public buildings having different degrees of damages. Some of such damage phenomena seemed new at least in the Chinese mainland and somehow educative.

The Lushan Gymnasium, a large-span space steel roof truss structure, has been investigated (Xiong et al. 2013). It was shown that the steel grid roof damaged heavily, which included buckling and crack of steel rod and support (see Fig. 8.7). Although the reasons for this kind of damage may be complicated which include

Fig. 8.6 Damage of the infilled wall in an office of the Lushan Forestry Bureau, picture from Gong et al. (2013)

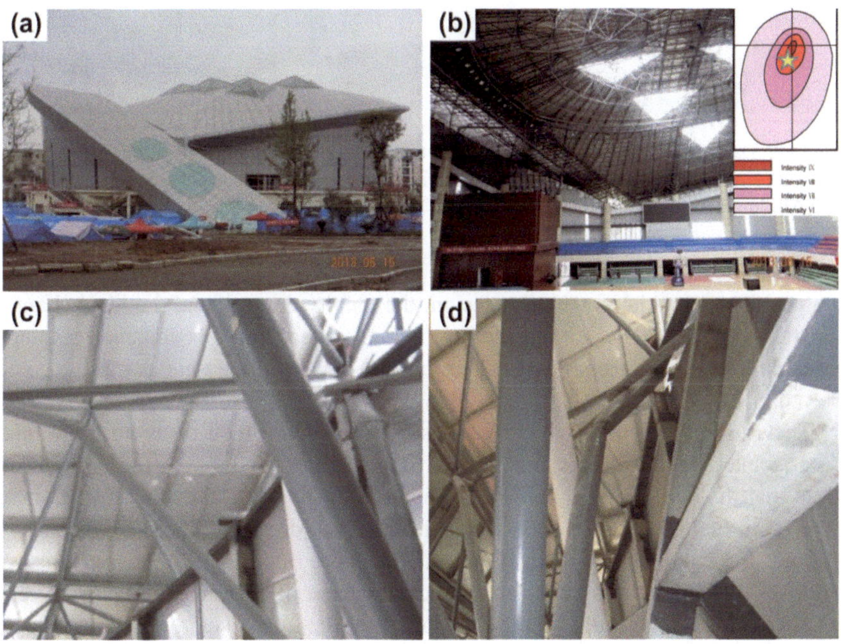

Fig. 8.7 Damage of the Lushan Gymnasium, built in 2012 (Xiong et al. 2013)

Fig. 8.8 Damage of a base-isolated structure: the Lushan Hospital. Investigation indicated that the main structure was basically undamaged, with slight detachments of the in-door decorations. The building came into use just 5 days after the Lushan earthquake

the problems associated with the design, the quality control, and the local characteristics of the seismic ground motion, among others, this phenomenon reminds researchers and engineers that steel structure might be not as safe as one expected before.

The Lushan Hospital, which is a 6-story steel-reinforced concrete frame and base-isolated building (Figs. 8.8, 8.9, 8.10), performed well during the earthquake. Note that this is the first time in Lushan and Wenchuan earthquake affected area, or even in mainland China, that a base-isolated building was tested by the strong ground shaking of a real-life earthquake. Due to the importance of hospital in saving lives during the earthquake emergency, this case of base-isolation plays a very positive role in the rescue and relief actions after the Lushan earthquake.

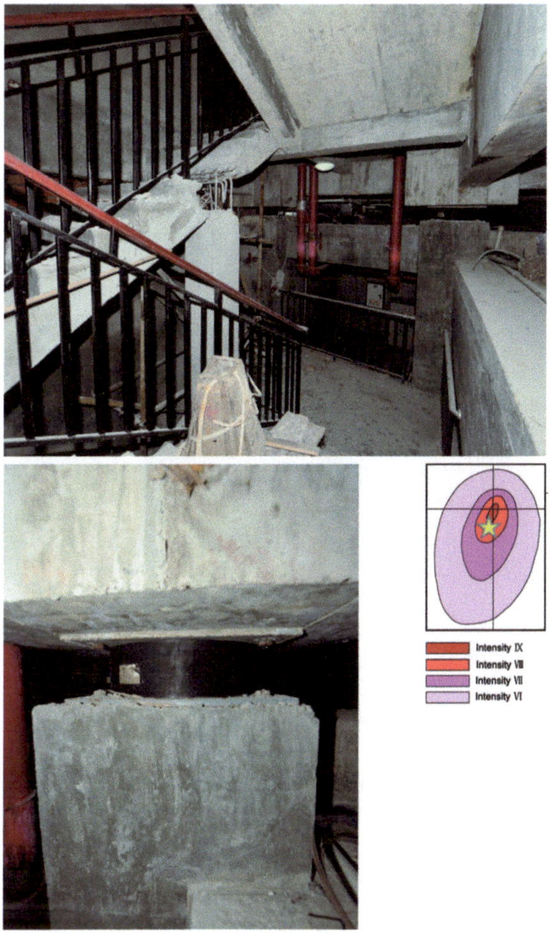

Fig. 8.9 Details of the base isolation in the Lushan Hospital

8.3 Lifeline Systems

Economic losses caused by an earthquake come from many factors, in which the destructions of the lifeline system, such as transportation, electricity, telecommunication, water supply system and water resources, and other related structures and/or equipments, play an essential role, not only due to the economic values of these systems but also due to the important functions of these systems in earthquake emergency. The Lushan earthquake caused some damages to the local lifeline systems, especially in the hardest-hit area. The destroyed systems included water and electricity supply systems, telecommunication stations and highways, which interfered the post-seismic relief works to some extent (Liu et al. 2013a, b, c; Ye et al. 2013).

Fig. 8.10 Damage of the Lushan Hospital. **a** Damage of the connection between the main building and the attachment. **b** Damage of the connection between the main building and the attached building

8.3.1 Transportation

Lushan and its nearby counties are in the mountainous area, where roads are the most fundamental and efficient transportation facilities. Earthquakes in this region usually caused landslides and rock falls, resulting in the blocks of the transportation and threatening the travelers along the roads. According to field investigation, in the hardest-hit area, almost all the roads were affected. As a result, transportations near Ganzi County along state road G318, near Lushi along province road S211, near Feixianguan along province road S210 were totally cut off (Fig. 8.11). Nearly 300 bridges damaged in different degrees, among which nearly 80 % were the arch bridges. Bridges named Baosheng (Fig. 8.12), Yuxi, Longmen, Lianghekou, Jinmenguan, Fandi, and Daduhe were damaged heavily.

Fig. 8.11 Road destroyed by a giant stone at province road S210 near Feixianguan Town

Fig. 8.12 The Baosheng Bridge destroyed by a huge stone of landslide at Baosheng Village, Lushan

Fig. 8.13 Damage of Tianquan transformer substation. **a** Leaking of pipe of transformer, No. 1 main transformer substation. **b** Damage of sulfur hexafluoride breaker

8.3.2 Electricity, Telecommunication, Water Supply and Sewage

Electricity grid is totally destructed in Baoxing, Lushan and Tianquan County during the earthquake. Besides, Yucheng District in Ya'an City experienced power off. Ya'an Electricity Company, which located near the epicenter, was affected tremendously by the earthquake. Two 100 kV Transformer Substation stopped functioning, with one of which irreparable. Leakage, displacement and deformation of key equipments resulted in the power off (Fig. 8.13).

Tele-communication system was also affected by this earthquake. With the effect mainly including base station damage (Fig. 8.14), equipment damage (Fig. 8.15), telecommunication pole damage and optical cable damage. In quake hit area, 724 of 2,786 base stations were cut off, which resulted in the break of telecommunications within 16 counties.

Fig. 8.14 Collapse and electricity power cut in a base station of China Unicom Sichuan Branch, Longmen Town, Lushan

Fig. 8.15 Damage in the central base room of China Unicom Sichuan branch at Tianquan County

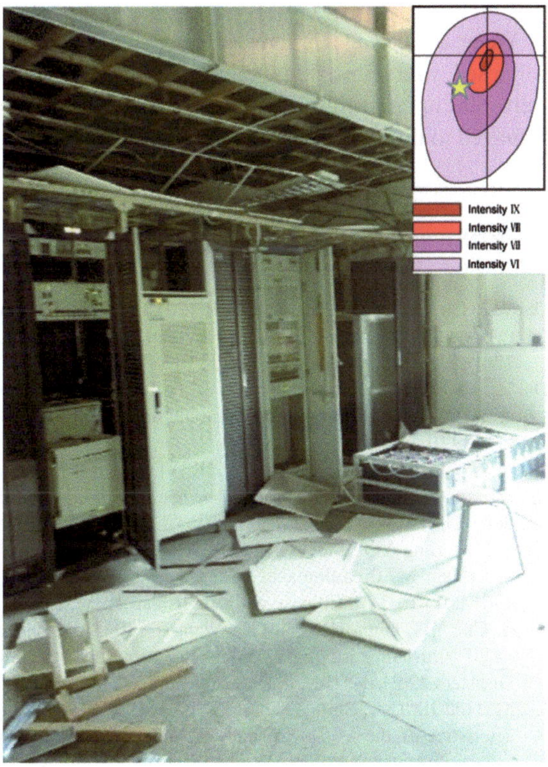

Fig. 8.16 Damage of pipe of Ren Yi Water Company at Tianquan County

Water and sewage system were impacted as well. Many parts of the river banks collapsed, which involved Qingyi River, Baoxing River and Lushanhe River. There were 500 or more reservoirs, 300 hydropower stations, and about 500 km dike damaged to different extents (Fig. 8.16).

References

Dai JW, Qu Z, Zhang CX, Weng XR (2013) Preliminary investigation of seismic damage to two steel space structure during the 2013 Lushan earthquake. Earthq Eng Eng Vibration 12(3):497–500. doi:10.1007/s11803-013-0189-6

Dai ZJ, Li XJ (2013) An explanation of the large PGA value of 2013 M_S7.0 Lushan earthquake at 51BXD station through topographic analysis. Earthq Sci 26(3–4):199–205. doi:10.1007/s11589-013-0043-y

Gao YW, Jin B, Dai JW, Chen HF (2013) Damage to rural wooden structures in M7.0 Lushan earthquake and reinforcement suggestions. J Earthq Eng Eng Vibration 33(3):48–52. doi:10.11810/100-1301.20130306 (in Chinese with English abstract)

Gong MS, Yang YQ, Xie LL (2013) Seismic damage to reinforced concrete frame buildings in Lushan M7.0 earthquake. J Earthq Eng Eng Vibration 33(3):20–26. doi:10.11810/1000-1301.20130303 (in Chinese with English abstract)

Jin B, Gao YW, Chen HF, Liu GQ, Wang T (2013) Primary damage analysis and study of the "4.20" Lushan earthquake on single-story factory buildings with brick column. J Nat Disasters 22(3):1–8 (in Chinese with English abstract)

Li ZQ, Hou JS, Li Y, Su GW, Jiang LX, Sun BT, Yuan YF, Lin JQ (2013) Analysis on the characteristics of the M_S7.0 Lushan, Sichuan Province, earthquake hazard on April 20, 2013.

Seismol Geol 35:398–410. doi:10.3969/j.issn.0253-4967.2013.02.018 (in Chinese with English abstract)

Liu JL, Lin JQ, Liu RS, Guo ED, Ye F, Yan DQ (2013a) Investigation and analysis of seismic damage of highway system in Lushan "4.20" earthquake. J Nat Disasters 22(3):18–23 (in Chinese with English abstract)

Liu JL, Lin JQ, Liu RS, Guo ED, Ye F, Yan DQ (2013b) Seismic damage investigation and analysis of the communication system in Lushan M_S7.0 earthquake. J Nat Disasters 22(5):91–97 (in Chinese with English abstract)

Liu RS, Liu JL, Yan DQ, Ye F, Lin JQ, Li J (2013c) Seismic damage investigation and analysis of electric power system in Lushan M_S7.0 earthquake. J Nat Disasters 22(5):83–90 (in Chinese with English abstract)

Qu Z, Zhong JR, Sun JJ (2013) Seismic damage to masonry structures in M7.0 Lushan earthquake. J Earthq Eng Eng Vibration 33(2):27–35. doi:10.11810/1000-1301.20130304 (in Chinese with English abstract)

Sun BT, Wang MZ, Yan PL, Zhang HY, Chen XZ (2013) Damage characteristics and seismic analysis of single-story brick bent frame column industrial buildings in Lushan M_S7.0 earthquake. J Earthq Eng Eng Vibration 33(3):1–8. doi:10.11810/1000-1301.20130301 (in Chinese with English abstract)

Wang YM, Xing LH, Xu WX (2013a) Seismic damage and damage enlightenment of medical buildings in Lushan M_S7.0 earthquake. J Earthq Eng Eng Vibration 33(4):44–53. doi:10.11810/1000-1301.20130406 (in Chinese with English abstract)

Wang T, Lei YD, Zhang YQ, Kong ZA (2013b) Seismic damage analysis of buildings in Lushan earthquake. J Earthq Eng Eng Vibration 33(2):36–47. doi:10.118101/1000-1301.20130305 (in Chinese with English abstract)

Wang T, Zhang YQ, Jin B, Kong ZA, Lei YD (2013c) Seismic damage to masonry residential buildings in Lushan M_S7.0 earthquake. J Earthq Eng Eng Vibration 33(3):9–19. doi:10.11810/1000-1301.20130302 (in Chinese with English abstract)

Wen RZ, Ren YF, Qi WH, Lu T, Yang ZY, Shan ZD, Wang YL (2013) Maximum acceleration recording from Lushan earthquake on April 20. J Southwest Jiaotong Univ 48(5):783–791

Xie J, Li X, Wen Z, Wu C (2014) Characteristics of near-source vertical and horizontal strong ground motion from the 20 April 2013 M_W6.8 Lushan earthquake in China. Seism Res Lett 85:23–33. doi:10.1785/0220130121

Xiong LH, Lan RQ, Wang YM, Tian XM, Feng B (2013) Earthquake damage investigation of structures in 7.0 Lushan strong earthquake. J Earthq Eng Eng Vibration 33(4):35–43. doi:10.11810/1000-1301.20130405 (in Chinese with English abstract)

Xu C, Chen B, Li XJ, Chen XL, Lü HS, Geng F, Wen ZP (2013) Seismic performance characteristics and damage of buildings in Lushan M_S7.0 Earthquake. Acta Seismol Sin 24:749–758. doi:10.3969/j.issn.0253-3782.2013.05.013 (in Chinese with English abstract)

Ye F, Guo ED, Liu JL, Liu RS, Lin JQ, Liu Z (2013) Seismic damage investigation and analysis of gas system in Lushan M_S7.0 earthquake. J Nat Disasters 22(5):77–82 (in Chinese with English abstract)

Zhou TG, Qian XB, Zhang BB (2013) Investigation and analysis of damage to rural houses in Lushan earthquake. J Earthq Eng Eng Vibration 33(3):53–58. doi:10.11810/1000-1301.20130307 (in Chinese with English abstract)

Chapter 9
Time-Dependent Seismic Hazard of the Southern Longmenshan Fault Zone and the Sichuan-Yunnan Region: Earthquake Predictability and Its Limit

Abstract Time-dependent seismic hazard is one of the critical issues for the social sustainability and one of the challenging scientific problems in earthquake science. For the southern Longmenshan fault zone as well as the Sichuan-Yunnan region, lessons and experiences of the Lushan earthquake provide rich raw materials for further scientific investigation, albeit still with a long distance to the systematic and practice-oriented understandings.

Keywords Earthquake Predictability · Seismic Hazard · The southern Longmenshan fault zone · Reverse tracing of precursors (RTP)

9.1 The Lushan Earthquake in a CSEP Testing Region

In recent years, study on earthquake predictability has gradually moved from a 'silver-bullet approach' to a 'brick-by-brick approach' (Jordan 2006). One of the representative projects is the Collaboratory for the Study of Earthquake Predictability (CSEP) project which aims at the establishment of an adequate infrastructure for conducting prospective prediction experiments under rigorous, controlled conditions and evaluating them using accepted criteria specified in advance.[1] Years of efforts have lead to a virtual, distributed laboratory that can support a wide range of scientific prediction experiments in multiple regional or global natural laboratories (Jordan 2006). Regional experiments involving time-independent and time-dependent models are now underway in California, Italy, Japan, New Zealand, Northwest Pacific, Southwest Pacific, and other places. The

[1] http://www.cseptesting.org/

This chapter is written by Jiang CS and Wu ZL (Institute of Geophysics, CEA) and Li YC (Earthquake Administration of Jiangsu Province).

DOI: 10.1007/978-981-4585-15-6_9, © The Author(s) 2014

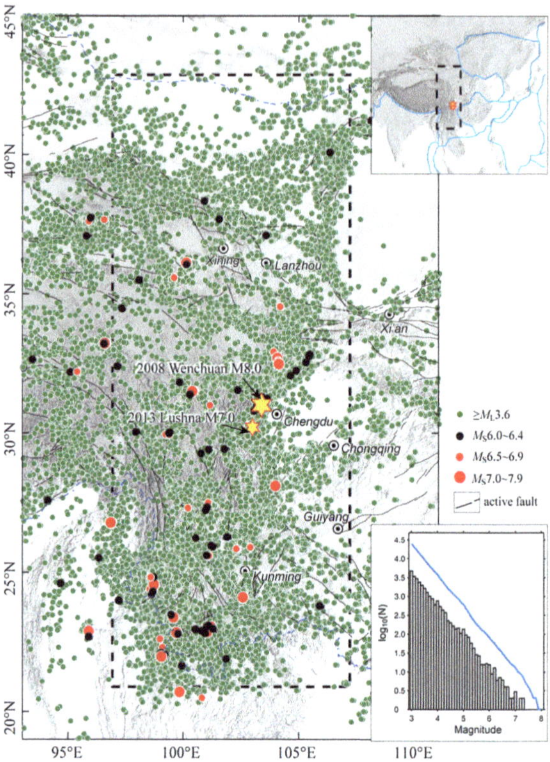

Fig. 9.1 The CSEP testing region of the central China north-south seismic zone, with the distribution of earthquakes above $M_L3.5$ since 1970, highlighting the 2008 Wenchuan earthquake and the 2013 Lushan earthquake. Indexing figure is to the *top right*, with plate boundaries (Bird 2003) shown in blue. To the *bottom right* is the frequency-magnitude distribution of the recorded earthquakes since 1970, from which it may be seen that the completeness magnitude is down to $M_L3.0$

central China north-south seismic zone (Fig. 9.1) is a new comer but an interesting and to much extent challenging testing region.

This testing is special firstly due to its intense seismic activity. Chinese historical literatures recorded several magnitude 8 earthquakes in this region, including the 1654 South of Tianshui M8 earthquake, the 1739 Yinchuan-Pingluo M8 earthquake, the 1833 Songming M8 earthquake, the 1879 South of Wudu M8 earthquake, the 1920 Haiyuan M8½ earthquake, the 1927 Gulang M8 earthquake, and the 2008 Wenchuan M8.0 earthquake.

Complex tectonics with a unified geodynamic background is another feature of this testing region. In the perspective of geology, the central China north-south seismic zone lies in between the Tibetan Plateau, the Ordos block, the Sichuan basin, and the South China block (Zhang et al. 2003). This belt is also marked by sharp and systematic variation of crust thickness from 30–46 km in the east to 46–74 km in the west, reflecting the rising and spreading of the Tibetan plateau with the collision between the India plate and the Eurasia plate (Li et al. 2006).

Seismological observation in this region has established a homogeneous monitoring capability of completeness magnitude $M_L3.0$ (Mignan et al. 2013), providing the background for the systematic test of the predictive models. This completeness magnitude is relatively higher than those in other testing regions with good observational facilities. But considering the magnitude of the 'target earthquakes' (major to great ones) and the time span of the continuous observation, this region is still good for the testing of the predictive algorithms.

Considering the CSEP testing region of the central China north-south seismic zone as the background, the review of the works related to the time-dependent seismic hazard and the precursor-like anomalies may provide some fresh and vivid, although not systematic, even not rigorous either, raw materials, and probably heuristic clues.

9.2 Predictive Assessments of Time-Dependent Seismic Hazard Priori to the Lushan Earthquake

China has established a system for the testing of the predictive assessments of time-dependent seismic hazard since the early 1970s. Although the practical successes have been still limited and the scientific bases were subject to debate, this system is advantageous in its real forward prediction nature and long-lasting feature, which are all the important issues in the study of earthquake predictability. Here we use the terminology 'predictive assessment of time-dependent seismic hazard', rather than simply 'forecast' or 'prediction' since in China, the word 'Dizhen (earthquake) Yuce (forecast)/Yubao (prediction)' is used with a much wider range than that in English. In most of the cases it is much closer to the time-dependent seismic hazard, rather than to 'earthquake prediction' in English. Figure 9.2 summarizes such predictive assessments for future study, which is to be discussed as follow.

At the intermediate-to-long-term time scale, according to the conclusions of the CEA's Research Group of "Researches on Earthquake Risk Regions and Losses Prediction of China Continent During from 2006 to 2020" (2007), the southern segment of the Xianshuihe fault zone and the middle-to-southern segment of the Longmenshan fault zone were delineated as a potentially seismic risky region for the period from 2006–2020, which is shown to be correct since the 2008 Wenchuan earthquake nucleated in the middle segment of the Longmenshan fault zone and the 2013 Lushan earthquake ruptured in the southern segment of the Longmenshan fault zone. At a shorter time scale of decade, after the Wenchuan earthquake, the CEA's Working Group of M7 (2012) concluded that the southern segment of the Longmenshan fault zone was subject to an earthquake with magnitude no less than 7 (In their expression, the degree of seismic hazard is of the second class as comparing to the 'seismic risky regions').

The southern segment of the Longmenshan fault zone has been keeping quite since the 2008 Wencuan M8.0 earthquake, which caused the concern of a strong aftershock, or another independent mainshock. Shortly after the Wenchuan earthquake, based on the rupture process inversion result, Yun-tai Chen warned that the hazard of a strong aftershock has to be accounted for (see, Chen et al. 2013a). Similar concerns and debates were also among the experts in different research fields. In the discussion on the site selection of the Wenchuan-earthquake Fault Scientific Drilling (WFSD), one of the proposals subject to debate was to have a drilling site in the southern segment of the Longmenshan fault zone. But this proposal was not accepted by the expert panel.

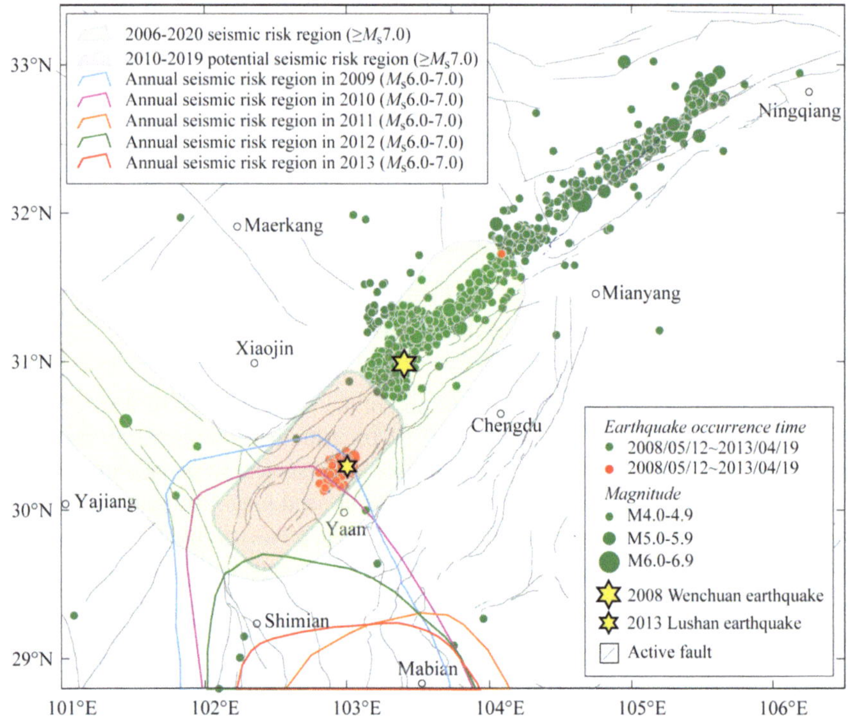

Fig. 9.2 Predictive information at different time-scales, mainly from the internal report of the China Earthquake Administration (CEA)

At the annual time scale, the predictive statements are somehow educative. Basically the 'Annual Consultation on the Likelihood of Earthquakes' is observation-based, but the final conclusion is dependent on panel discussion, which relies to much extent on the empirical experiences of the experts. In the year of 2009 and 2010, a region with higher probability of earthquakes covered the epicenter of the Lushan M7.0 earthquake. However, in 2011, 2012, and 2013, the Lushan epicenter moved out of the annual seismic risky region.[2] Apparently in 2009 and 2010, the concern of a strong aftershock in the southern Longmenshan fault zone plays an important role in the Annual Consultation.

At shorter time scale the predictive statements kept controversial. The Survey Engineering Institute of Sichuan Earthquake Administration sent a report to the Earthquake Administration of Sichuan Province on February 22, 2013, predicting a $M_S6.0 \sim 6.9$ earthquake in a ellipse region covering the south Longmenshan fault zone, the north Anninghe fault zone, and the south Xianshuihe fault zone, in a

[2] Working Group of the Scientific Review and Reflection of the Prediction of the 2013 Lushan M7.0 Earthquake (group leader Liu GP, 2013) Internal Report of the Scientific Review and Reflection of the Prediction of the 2013 Lushan M7.0 Earthquake. Department for Earthquake Monitoring and Prediction, China Earthquake Administration, Beijing (in Chinese).

~75 days forecasting window. A piece of news reported this short-term forecast after the Lushan earthquake and caused some discussions in the society.[3] The basis of the prediction is the leveling measurement along the Xianshuihe fault zone, plus some considerations of gravity field variation. The prediction was not agreed with in the Earthquake Administration of Sichuan Province and the China Earthquake Networks Center (CENC). A retrospective investigation shows that before this prediction there were some predictions for approximately the same region, with false-alarms. It turns out that most likely what was observed is still an intermediate-term anomaly rather than a short-term anomaly if the observation could be confirmed. Moreover, since the measurement was basically along the Xianshuihe fault zone, but the prediction region included the Xianshuihe, the Anninghe, and the Longmenshan fault zone, the physical picture of the prediction is worth further discussion.

9.3 Forecast of the Lushan Earthquake? Forward Predictive Statements and Retrospective Case Studies

9.3.1 Predictive Statements in Academic Publications

Century scale: Comparing the accumulated moment on a fault over a period with the seismic moment release during the period, one may estimate the moment deficit available for future earthquakes. This physical picture is simply to consider seismic activity as a process of the accumulation and release of tectonic stress. After the Wenchuan M_S8.0 earthquake, Wang et al. (2010) re-assessed the seismic hazard of the Lomgmenshan fault zone by using this moment balancing approach, and found that the largest moment deficit was still on the south segment of the Longmenshan fault zone after the Wenchuan earthquake. This clearly indicated seismic hazard within the next half century with the maximum magnitude M_W7.7 (if the whole segment ruptures).

Decades scale: Calculations of the perturbation of the Coulomb failure stress (CFS) by the 2008 Wenchuan earthquake, although being different in details, are almost consistent in indicating that the southern Longmenshan fault zone is a place with increased CFS (Parsons et al. 2008; Toda et al. 2008; Shan et al. 2009; Wan et al. 2009; Xie et al. 2010), which clearly implies the seismic hazard at the decades time scale. Similar information is from the analysis of seismicity to deduce the stress level using the Gutenberg-Richter *b*-value. The result of Yi et al. (2013) indicates an increase of stress level in the region of Tianquan, Lushan, Luding and north of Baosing after the 2008 Wenchuan earthquake. After the Wenchuan earthquake, a retrospective study was conducted by Jiang and Zhuang (2010) to investigate whether long-term seismicity anomalies exist before the Wenchuan earthquake, and to evaluate the current potential risks of strong earthquakes in the Sichuan-Yunnan region. They found that there exists a long-term and large area of low 'clustering ratio' in the middle-to-south segment of the Longmenshan

[3] http://www.infzm.com/content/90403

fault zone, implying that this region had been in a state of stress barrier even before the Wenchuan earthquake.

Years to annual scale: Qin et al. (2013) reported their borehole stress measurement results in 2003, 2008, and 2010, in 4 boreholes located in Baoxing and Kangding, and found that the stress level was increased after the 2008 Wenchuan earthquake, and the maximum horizontal principal stress have reached the lower limits of fault activation in the south-west segment of the Longmenshan fault zone. There were even discussions on the 'scenario rupture' of the southern Longmenshan fault zone and the design of a system of 'monitoring and modeling for prediction', albeit without clear indication of the time scale (Wu et al. 2010).

9.3.2 Retrospective Case Reports of the Precursor-Like Anomalies

After the Lushan earthquake, there were some reports of retrospective case studies of the precursor-like anomalies observed, including borehole strain (Chi 2013; Chi et al. 2013; Qiu et al. 2013), gravity anomaly (Zhu et al. 2013; Liang et al. 2013; Jiang et al. 2014b), deformation anomaly (Niu et al. 2013), variation of seismicity (Rong and Li 2013), apparent stress (Gong et al. 2013) and velocity of seismic waves (Wang et al. 2014), the Outgoing Long-wave Radiation (OLR, Guo et al. 2013), anomalies in the ionosphere (Jiang et al. 2013a), cloud (Wu et al. 2013), infrasonic signals (Pan et al. 2013), and ground temperature (Chen et al. 2013b; Zhang et al. 2013; Xu and Xu 2013), as well as in geomagnetic and geoeletric field (An et al. 2013; Ma et al. 2013).

Figure 9.3 shows the related reports together with their spatial and temporal scales. It is too early to get definite conclusions using the limited data set, although in the figure it seems that the characteristic distance and time are reversely proportional to each other, that is, near to the time of the earthquake (when the time scale is getting smaller) the spatial range of the anomalies is larger and larger. This, if could be confirmed, might be related to the criticality associated with the earthquake preparation process (Main 1996; Rundle et al. 1997) in which the correlation length becomes infinite approaching the critical point of the system.

9.4 Discussion on the Future: A RTP-Like Analysis Focusing on the Southern Longmenshan Fault Zone

Reviewing the predictive analyses, either forward or retrospective, is to think of the future. Facing to the Xianshuihe fault along which the last strong earthquake has been 33 years ago, the southern Longmenshan fault zone along which there are apparently two 'seismic gaps' left after the Wenchuan earthquake and the Lushan earthquake, and the Anninghe fault zone which forms the 'triple junction'

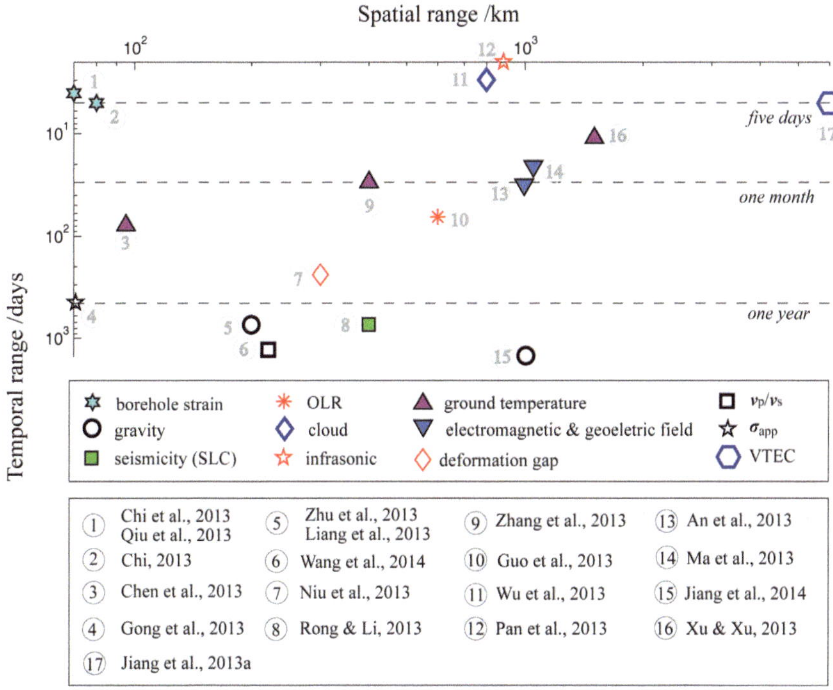

Fig. 9.3 Retrospective reports of the predictive anomalies, in the coordinate of space and time range. From the figure it seems that, except the borehole strain anomaly ① and ② which shows only the result of the site near the epicenter, a weak trend exists that the characteristic spatial range and time range (in the logarithm coordinate) are reversely proportional to each other

with the former two fault zones, we find that the knowledge of earthquakes at the present time is still insufficient to provide a reliable assessment of seismic hazard in the near future. However, we may try to use a RTP-like approach to review, at different time scales, the seismic hazard.

The original reverse tracing of precursors (RTP) approach was proposed by Keilis-Borok et al. (2004) who connected the short-term seismicity patterns with intermediate-term ones. These patterns are analyzed in the reverse order of their appearance: short-term patterns are analyzed first, although they emerge later. This approach has a natural Earth-specific conceptual explanation that strong earthquake is a result of a long-lasting large-scale process whose different stages involve different parts of the fault network. Physically, RTP approach tries to identify a rare small-scale phenomenon that carries a memory of the larger scale history of the system. Although subject to discussion and debate (Zechar and Zhuang 2010), this idea can be borrowed to the present problem that considering the being concerned 'seismic gaps' along the south Longmenshan fault zone, anomalies at longer time scales oriented at these 'gaps' be considered.

After the Lushan earthquake, Liu et al. (2014) recalculated the moment deficit and the Coulomb failure stress changes, and concluded that the southern

Table 9.1 Coulomb failure stress (CFS) changes of the 2013 Lushan earthquake and the 2008 Wenchuan earthquake

Authors	Stress changes considered	Models	ΔCFS in Lushan EQ epicenter (MPa)	Relation to Wenchuan EQ	ΔCFS increased region
Shan et al. 2013	Co-seismic and post-seismic	Multi-layered lithospheric	0.037 ~ 0.0113	Triggered	Daofu-Kangding segment in XSH; Southern segment of LMS
Miao and Zhu 2013	Co-seismic and post-seismic	Homogeneous elastic half-space	0.012	Triggered	South-east segment of XSH; Middle segment of LMS
Dong et al. 2013	Co-seismic of Lushan EQ.	Elastic/viscoelastic layered half space	–	–	Daofu-Kangding segment in XSH
Liu et al. 2014	Co-seismic and post-seismic	Finite element model	0.03 ~ 0.3	Partly triggered	XSH; Southern segment of LMS
Jia et al. 2014	Co-seismic and post-seismic	Elastic/viscoelastic layered half space	0.049	Triggered	–

Note: *LMS*—Longmenshan fault; *XSH*—Xianshuihe fault

Longmenshan fault zone is still having the moment deficit for producing another M7 earthquake. Meanwhile, CFS change indicates that both the southern Longmenshan fault zone and the south Xianshuihe fault zone are in the region with increased CFS. However for different models and different considerations, the results of the CFS change (Shan et al. 2013; Miao and Zhu 2013; Dong et al. 2013; Liu et al. 2014; Jia et al. 2014) are model dependent, as shown in Table 9.1. On the other hand, however, majority of the models indicate an increase of CFS along the south Xianshuihe and the south Longmenshan fault zone, which is consistent with the geodynamic intuition.

The PI forecast, which applies the concepts of statistical physics of complex systems to the analysis of seismic activity (Rundle et al. 2002), provides a picture of 5-year time-scale 'hotspots' (Holiday et al. 2005). Jiang and Wu (2008) conducted a retrospective forecast test of the PI algorithm for the Sichuan-Yunnan region. A similar retrospective case study of the Lushan earthquake was conducted by Jiang et al. (2013b). Remarkably, the distribution of the 'PI hotspots' indicates that the southern Longmenshan fault zone still has a higher seismic probability. Based on the time-space epidemic-type aftershock sequence (ETAS) model and stochastic declustering method, the 'clustering ratio' (Zhuang et al. 2005) mapped by Jiang et al. (2014b) enhances the seismic probability of the southern Longmenshan fault zone. Seen at different time scales, therefore, the seismic hazard of the southern Longmenshan fault zone in near future is still in need of close-in monitoring.

Last but not the least, the concern of the time-dependent seismic hazard along the south Longmenshan fault zone is, to much extent, based on the concept of 'seismic gap'. Recent works also used this picture to analyze the long-term seismic hazard in west China (Qu et al. 2010). Historical evolution of this concept marked the timeliness starting from the 1906 San Francisco earthquake, about which Reid (1910) proposed the elastic rebound model. Imamura (1928) documented historical earthquakes in the Nankai trough, southwest Japan, used the regularity of the occurrence of large earthquakes, and even forecasted the large earthquakes in this area which was tested by the two large events with $M \approx 8$ in 1944 and 1946. The presently well-known 'seismic gap' concept may be from Fedotov (1965). Debates on the results of the 'seismic gaps' in the perspective of statistical test (Kagan and Jackson 1995) made this problem more complicated. And how far away can we be guided by this theoretical concept is still an open question.

References

An ZH, Du XB, Tan DC, Fan YY, Liu J, Cui TF (2013) Study on the geo-electric field variation of Sichuan Lushan M_S7.0 and Wenchuan M_S8.0 earthquake. Chin J Geophys 56:3868–3876. doi:10.6038/cjg20131128 (in Chinese with English abstract)

Chen SY, Liu PX, Liu LQ, Ma J (2013a) A phenomenon of ground temperature change prior to Lushan earthquake observed in Kangding. Seismol Geol 35:634–640. doi:10.3969/j.issn.0253-4967.2013.03.017 (in Chinese with English abstract)

Chen YT, Yang ZX, Zhang Y, Liu C (2013b) From 2008 Wenchuan earthquake to 2013 Lushan earthquake. Scientia Sinica Terrae 43:1064–1072 (in Chinese)

Chi SL (2013) Strain anomalies before Wenchuan and Lushan earthquakes recorded by component before strainmeter. Sci Technol Rev 31(12):27–30. doi:10.3981/j.issn.1000-7857.2013.12.004 (in Chinese with English abstract)

Chi SL, Liu Q, Chi Y, Deng T, Liao CW, Yang G, Zhang GP, Chen J (2013) Borehole strain anomalies before the 20 April 2013 Lushan M_S7.0 earthquake. Acta Seismol Sin 35:296–303. doi:10.3969/j.issn.0253-3782.2013.03.002 (in Chinese with English abstract)

Dong PY, Cheng HH, Zeng XF, Shi YL (2013) Stress changes on major faults by M_S7.0 Lushan earthquake. Sci Technol Rev 31(12): 19–22. doi: 10.3981/j.issn.1000-7857.2013.12.002 (in Chinese with English abstract)

Fedotov SA (1965) Regularities of the distribution of strong earthquakes in Kamchatka, the Kurile Islands and northeastern Japan. Tr Inst Fiz Zemli, Akad Nauk SSSR 36(203):66–93 (in Russian)

Gong Y, Long F, Yi GX, Zhao M, Qiao HZ, Wang YX (2013) Study of apparent stress on the Southern segment of Longmenshan fault zone around the Lushan earthquake. Technol Earthq Disaster Prev 8(4):389–396 (in Chinese with English abstract)

Guo X, Zhang YS, Wei CX, Zhong MJ, Zhang X (2013) OLR anomalies for the Lushan M_S7.0 earthquake. Acta Seismol Sin 35:731–737. doi:10.3969/j.issn.0253-3782.2013.05.011 (in Chinese with English abstract)

Holliday JR, Nanjo KZ, Tiampo KF, Rundle JB, Turcotte DL (2005) Earthquake forecasting and its verification. Nonlinear Proc Geophys 12:965–977

Imamura A (1928) On the seismic activity of central Japan. Japan J Astron Geophys 6:119–137

Jia K, Zhou SY, Zhuang JC, Jiang CS (2014) Possibility of the independence between the 2013 Lushan earthquake and the 2008 Wenchuan earthquake on Longmen Shan fault, Sichuan. Seism Res Lett 85:60–67. doi:10.1785/0220130115

Jiang CS, Wu ZL (2008) Retrospective forecasting test of a statistical physics model for earthquakes in Sichuan-Yunnan region. Sci Chin Ser D: Earth Sci 51: 1401–1410

Jiang CS, Zhuang JC (2010) Evaluation of background seismicity and potential source zones of strong earthquakes in the Sichuan-Yunan region based on the space-time ETAS model. Chin J Geophys 53:305–317. doi:10.3969/j.issn.0001-5733.2010.02.001 (in Chinese with English abstract)

Jiang CS, Wu ZL, Zhuang JC (2014b) Clustering ratio method for the intermediate-term and long-term earthquake forecasting: a case study in the North-south Seismic Belt. Manuscript (in Chinese with English abstract)

Jiang H, Wu ZL, Ma TF, Jiang CS (2013b) Retrospective test of the PI forecast: case study of the April 20 2013, Lushan, Sichuan, China, M_S7.0 earthquake. Physics 42:334–340. doi:10.7693/wl20130504 (in Chinese with English abstract)

Jiang L, Li DQ, Xu ZP, Yang LP, Xiong W, Xu SQ (2014a) Characteristics of gravity field before the Lushan M_S7.0 earthquake based on time variable gravity data from GRACE satellite. Acta Seismol Sin 36(1):84–94 (in Chinese with English abstract)

Jiang WP, Ma YF, Liu HF, Deng LS, Zhou XH (2013a) Investigation of Lushan earthquake ionosphere VTEC anomalies based on GPS data. Earthq Sci 26(3–4):259–265

Jordan TH (2006) Earthquake predictability, brick by brick. Seismol Res Lett 77:3–6

Kagan YY, Jackson DD (1995) New seismic gap hypothesis: five years after. J Geophys Res 100:3943–3959. doi:10.1029/94JB03014

Keilis-Borok V, Shebalin P, Gabrielov A, Turcotte D (2004) Reverse tracing of short-term earthquake precursors. Phys Earth Planet Inter 145:75–85

Li SL, Mooney WD, Fan JC (2006) Crustal structure of mainland China from deep seismic sounding data. Tectonophysics 420:239–252

Liang WF, Liu F, Xu YM, Zhu YQ, Guo SS, Zhao YF, Liu L (2013) Gravity observation in the eastern margin of Qinghai-Tibetan plateau and reflect to the Lushan M7.0 earthquake. Chin Earthq Eng J 2:266–271. doi:10.3969/j.issn.1000-0844.2013.02.0266 (in Chinese with English abstract)

Liu M, Luo G, Wang H (2014) The 2013 Lushan earthquake in China tests hazard assessments. Seism Res Lett 85:40–43. doi:10.1785/0220130117

Ma QZ, Fang GQ, Li W, Zhou JN (2013) Electromagnetic anomalies before the 2013 Lushan M_S7.0 earthquake. Acta Seismol Sin 35:717–730. doi:10.3969/j.issn.0253-3782.2013.05.010 (in Chinese with English abstract)

Main I (1996) Statitical physics, seismogenesis, and seismic hazard. Rev Geophys 34:433–462

Miao M, Zhu SB (2013) The static Coulomb stress change of the 2013 Lushan M_S7.0 earthquake and its impact on the spatial distribution of aftershocks. Acta Seismol Sin 35:619–631. doi:10.3969/j.issn.0253-3782.2013.05.001 (in Chinese with English abstract)

Mignan A, Jiang CS, Zechar JD, Wiemer S, Wu ZL, Huang ZB (2013) Completeness of the mainland China earthquake catalog and implications for the setup of the China Earthquake Forecast Testing Center. Bull Seismol Soc Amer 103(2A):845–859

Niu AF, Zhang LK, Li Y, Yan Wei, Zhao J, Ji P (2013) On the weak deformation area prior to the Lushan M_S7.0 earthquake. Technol Earthq Disaster Prev 8(4): 341–350 (in Chinese with English abstract)

Pan LL, Zeng ZX, Wang J (2013) Precursory abnormal infrasonic signals predating the Lushan earthquake (M_S7.0) and Yushu earthquake (M_S5.2). Earth Sci Front 20:73–79 (in Chinese with English abstract)

Parsons T, Ji C, Kirby E (2008) Stress changes from the 2008 Wenchuan earthquake and increase hazard in the Sichuan basin. Nature 454:509–510

Qin XH, Chen QC, Tan CX, An QM, Wu ML, Feng CJ (2013) Analysis of current geostress state and seismic risk in southwest segment of Longmenshan fracture belt. Chin J Rock Mech Eng 32:2870–2876 (in Chinese with English abstract)

Qiu ZH, Yang G, Tang L, Guo YP, Zhang BH (2013) Abnormal strain changes observed by a borehole strainmeter at Guza Station before the M_S7.0 Lushan earthquake. J Geodesy Geodyn 4(3): 19–29. doi: 10.3724/SP.J.1246.2013.03019

Qu YJ, Wang HT, Wu CD, Feng JG, Chen YW, Li YZ, Wang X (2010) Analysis on statistic characteristics of seismic gap in Chinese mainland. Acta Seismol Sin 32:544–556. doi:10.3969/j.issn.0253-3782.2010.05.004 (in Chinese with English abstract)

Reid HF (1910) The California earthquake of April 18, 1906. The mechanics of the earthquake, vol. 2. Carnegie Institution of Washington, Washington

Research group of "Researches on earthquake risk regions and losses prediction of china continent during from 2006 to 2020" (2007) Researches on earthquake risk regions and losses prediction of china continent during from 2006 to 2020. Seismological Press, Beijing, p 225

Rong DL, Li YR (2013) The critical point behavior of seismic activities prior to Lushan M_S7.0 earthquake in 2013. Chin Earthq Eng J 2:252–256. doi:10.3969/j.issn.1000-0844.2013.02.0252 (in Chinese with English abstract)

Rundle JB, Gross S, Klein W, Ferguson C, Turcotte DL (1997) The statistical mechanics of earthquakes. Tectonophysics 277:147–164

Rundle JB, Tiampo KF, Klein W, Matins JSA (2002) Self-organization in leaky threshold systems: the influence of near mean field dynamics and its implications for earthquakes, neurobiology, and forecasting. Proc Natl Acad Sci USA 99(Suppl 1):2514–2521

Shan B, Xiong X, Zheng Y, Diao FQ (2009) Stress changes on major faults caused by M_W7.9 Wenchuan earthquake, May 12, 2008. Sci Chin Ser D: Earth Sci 52:593–601. doi:10.1007/s11430-009-0060-9 (in Chinese with English abstract)

Shan B, Xiong X, Zheng Y, Jin BK, Liu CL, Xie ZJ, Xu HZ (2013) Stress changes on major faults caused by 2013 Lushan earthquake, and its relationship with 2008 Wenchuan earthquake. Sci Chin: Earth Sci 56:1169–1176. doi:10.1007/s11430-013-4642-1

Toda S, Lin J, Meghraoui M, Stein RS (2008) 12 May 2008 $M = 7.9$ Wenchuan, China, earthquake calculated to increase failure stress and seismicity rate on three major fault sysytems. Geophys Res Lett 35:L17305. doi:10.1029/2008GL034903

Wan YG, Shen ZK, Sheng SZ, Xu XF (2009) The influence of 2008 Wenchuan earthquake on surrounding faults. Acta Seismol Sin 31:128–139 (in Chinese with English abstract)

Wang H, Liu M, Shen X, Liu J (2010) Balancing seismic moment in eastern Tibet: implications for the Great 2008 Wenchuan Earthquake. Tectonophysics 491(1–4):154–164. doi:10.1016/j.tecto.2009.09.022

Wang LY, Li YE, Zheng XY, Wang SW (2014) Temporal variation of v_p/v_s at single seismic station before the 2013 Lushan M_s7.0 earthquake. Acta Seismol Sin 36(1):42–58 (in Chinese with English abstract)

Working Group of M7 (2012) Study on the mid- to long-term potential of larger earthquakes on the chinese continent. Seismological Press, Beijing, p 321

Wu ZL, Li L, Liu GP, Jiang CS, Ma HS (2010) Monitoring and modeling the multi-time-scale seismic hazard of the southern Longmenshan fault: an experimental design of the monitoring and modeling for prediction system.T54B-01, AGU Fall Meeting 2010, San Francisco, http://adsabs.harvard.edu/abs/2010AGUFM.T54B..01W

Wu LX, Zheng S, Zhong XH, Qin K (2013) Satellite linear clouds anomaly before M_S7.0 Lushan earthquake. Sci Technol Rev 31(12): 23-26. doi: 10.3981/j.issn.1000-7857.2013.12.003 (in Chinese with English abstract)

Xie ZD, Zhu YQ, Lei XL, Yu HY, Hu XL (2010) Pattern of stress change and its effect on seismicity rate caused by M_s8.0 Wenchuan earthquake. Sci Chin: Earth Sci 40:688–698. doi:10.1007/s11430-010-4025-9 (in Chinese with English abstract)

Xu XD, Xu BH (2013) The infrared warming abnormality analysis of Lushan earthquake. Earth Sci Front 20(3):25–28 (in Chinese with English abstract)

Yi GX, Wen XZ, Xin H, Qiao HZ, Wang SW, Gong Y (2013) Stress state and major-earthquake risk on the southern segment of the Longmen Shan fault zone. Chin J Geophys 56:1112–1120. doi:10.6038/cjg20130407 (in Chinese with English abstract)

Zechar JD, Zhuang JC (2010) Risk and return: evaluating reverse tracing of precursors earthquake predictions. Geophys J Int 182:1319–1326

Zhang PZ, Deng QD, Zhang GM, Ma J, Gan WJ, Min W, Mao FY, Wang Q (2003) Active tectonic blocks and strong earthquakes in continental China. Sci Chin Ser D 33(Supp l): 13–24

Zhang X, Zhang YS, Wei CX, Tian XF, Tang Q, Gao J (2013) Analysis of thermal infrared anomaly before the Lushan M_S7.0 earthquake. Chin Earthq Eng J 35:272–277. doi:10.3969/j.issn.1000-0844.2013.02.0272 (in Chinese with English abstract)

Zhu YQ, Wen XZ, Sun HP, Guo SS, Zhao YF (2013) Gravity changes before the Lushan, Sichuan, $M_S = 7.0$ Earthquake of 2013. Chin J Geophys 56:1887–1894. doi:10.6038/cjg20130611 (in Chinese with English abstract)

Zhuang JC, Chang CP, Ogata Y, Chen YI (2005) A study on the background and clustering seismicity in the Taiwan region by using point process models. J Geophys Res 110: B05S18, doi: 10.1029/2004JB003157

Chapter 10
Concluding Remarks Without Conclusions: Earthquake and Disaster Phenomenology not yet Included in our Textbooks

Abstract Each earthquake teaches us on the understanding of earthquakes and the reduction of earthquake disaster risk. Earthquake and disaster phenomenology, although being empirical and to much extent in lack of systematic features, play an important role both in the advancement of earthquake science and technology and in the collaborative efforts for social sustainability.

Keywords Earthquake cases · Earthquake phenomenology · The Lushan earthquake

In the perspective of either the magnitude or the disaster, even if in China, the Lushan earthquake is not a 'special event'. The observation, although having already had significant progresses even if comparing to 5 years ago, was only of 'pass' score comparing to the observations in other places of the world, such as in Parkfield. On the other hand, however, the Lushan earthquake did have some fresh phenomenology which has not yet been fully discussed. The aim of this Brief is simply to record these phenomenology for further studies.

In Chap. 3 we discussed the preliminary results in regard to the seismo-tectonics of the Lushan earthquake. Although the large scale geodynamic background of this earthquake is clear, the association of the earthquake with the known/unknown active faults remains puzzle. What is more worth discussion might not be such an association itself. Rather how to understand the role of such 'missing earthquakes' in our seismic hazard assessment is one of the blind spots in the present framework of knowledge and needs careful discussion.

Chapter 4 provides the results of the aftershock monitoring and analysis. These results are important constraints on the properties of the mainshock. Although the seismological tools for analyzing each aftershock has been well developed, the fresh phenomenology lies in the assembly of all the aftershocks and the relation with the mainshock.

This chapter is written by Wu ZL and Jiang CS (Institute of Geophysics, CEA).

Z. Wu et al., *Earthquake Phenomenology from the Field*, SpringerBriefs in Earth Sciences, 111
DOI: 10.1007/978-981-4585-15-6_10, © The Author(s) 2014

Chapter 5 is somehow special for the Lushan earthquake as per its relation with the Wenchuan earthquake. The 'aftershock debate' is somehow interesting in seismology because, going to details, the definition of aftershocks itself has so many issues to be discussed. Another interesting aspect is cultural, since the Chinese translation 'Yuzhen' (where Yu = remained and zhen = quake) seems having more physical contents than the original word 'aftershocks'. Nevertheless more practical works could be done concerning aftershocks, based on our present scientific knowledge (see also Appendix II).

Chapter 6 reflects a new trend in our time of information that on one hand, there is an increasing public demand of the seismological information service provided after each earthquake. On the other hand, seismological agencies equipped by real-time seismic recordings and computerized data processing systems are also more and more capable for providing better and better seismological information to the society. What has not been fully discussed is how to understand the process of such service, that is, the production, maintenance, quality control, and update of the scientific products, in an systems engineering perspective.

Chapter 7 totally belongs to our time with internet and twitter/microblog. What have been discussed there are only three examples of the new problems, namely the transmission of earthquake information through microblogs, the on-going estimate of the final losses based on internet reports, and the evacuation video recordings for constraining the physical models of escape panic. They are all useful for our society, and are fresh phenomenology for 'citizen seismology'.

In Chap. 8 we discussed some localized and might-be new phenomenology shown in the Lushan earthquake: the 'wooden structure puzzle', the 'standing ruins', and the (unexpected) destruction of large-scale steel structures. We discussed the lifeline system in which more works could be considered when we seriously consider what is the 'line' and what will happen if such 'lines' form a 'network'. Also note that the Lushan earthquake is the first case in the Chinese mainland in which base-isolation experienced the strong ground shaking as large as 0.5 g.

In Chap. 9 we discussed a challenging issue in earthquake science related to earthquake predictability. Related to the concern of the time-dependent seismic hazard of the southern Longmenshan fault zone (where there were two 'seismic gaps' left after the 2008 Wenchuan earthquake and the 2013 Lushan earthquake) and the 'triple junction' (among the Xianshuihe, Anninghe, and Longmenshan fault zone), our knowledge is apparently not enough.

This Brief is to much extent based on the Field Investigation of the 2013 Lushan Earthquake organized by the China Earthquake Administration (CEA). In Chap. 2 we give a brief introduction to the Field Investigation. Although it is a convention for scientists and scientific institutions to carry out field investigation after an earthquake, and each earthquake has different features, the field investigation itself seems still in need of further discussion. Note that after each air accident the forensic-like investigation has its own technical guidelines. Discussing on the guidelines for earthquake field investigation may be of help for better understanding the phenomenology of earthquakes and earthquake disasters.

It turns out that for the Lushan earthquake, what we have been asked is more than what we have learnt, and 'what we have to learn to do we learn by doing'. To much extent, this is a manifestation that earthquake science still has a seemingly endless frontier.

Appendix 1
Studies on the Lushan Earthquake

This appendix summarizes the up-to-date studies on the Lushan earthquake, highlighting the works carried out by Chinese institutions and published in Chinese journals.

Shortly after the Lushan earthquake, several publications appeared either in Chinese or in English, or both, see Table A1.1. Academic journals in China published special issues, or special-issue-like collection of papers, on the Lushan earthquake. The journals include *Acta Seismologica Sinica* (bulletin of the Seismological Society of China, vol. 35, issue 5), *Journal of Chengdu University of Technology* (Science & Technology edition, vol. 40, issue 3), *Science & Technology Review* (vol. 31, issue 12, published by the China Association of Science and Technology), *Earth Science Frontiers* (vol. 20, issue 3, published by the China University of Geosciences, Beijing), *Seismology and Geology* (vol. 35, issues 2 and 3, published by the Institute of Geology, China Earthquake Administration, Beijing), *Science in China: Earth Science* (vol. 56, issue 7, published by the Chinese Academy of Sciences, Beijing), *Chinese Journal of Geophysics* (bulletin of the Chinese Geophysical Society, vol. 56, issue 4), and *Journal of Earthquake Engineering and Engineering Vibration* (vol. 33, issue 3, published by the Institute of Engineering Mechanics, China Earthquake Administration, Harbin), in Chinese; and *Geodesy and Geophysics* (vol. 4, issue 3, published by the Institute of Seismology, China Earthquake Administration, Wuhan), *Chinese Science Bulletin* (vol. 58, issue 28/29, published by the Chinese Academy of Sciences, Beijing), and *Earthquake Science* (vol. 26, issue 3/4, the English edition of *Acta Seismologica Sinica*), in English.

Appendix 1 is the contribution of Zhang SF and Jiang CS, with the helps of Gao S and Wu ZL.

Z. Wu et al., *Earthquake Phenomenology from the Field*, SpringerBriefs in Earth Sciences,
DOI: 10.1007/978-981-4585-15-6,

Table A1.1 Publications on the Lushan earthquake

Topic	Authors
Mainshock: focal mechanism	Chen WW et al. 2013; Hu XG and Jiang Y 2013; Lin XD et al. 2013; Liu J et al. 2013; Xie ZJ et al. 2013; Zeng XF et al. 2013a, b
Mainshock: magnitude	Liu RF et al. 2013
Mainshock: rupture process	Hao JL et al. 2013; Liu CL et al. 2013; Sun XD et al. 2013; Wang WM et al. 2013; Xie ZJ et al. 2013; Xu LS et al. 2013; Xu Y and Shao WL 2013; Zhang LF et al. 2013; Zhang Y et al. 2013; Zhao CP et al. 2013; Jin MP et al. 2014; Zhao X et al. 2014; Zhang Y et al. 2014
Co-seismic variation: deformation	Du YJ et al. 2013; Liu G et al. 2013; Tan HB et al. 2013; Tang L and Jing Y (2013); Wang K et al. 2013; Wu YQ et al. 2013
Co-seismic variation: gravity	Wei J et al. 2013
Co-seismic variation: ionosphere	Cai H and Zhao GQ 2013; Zhou YY et al. 2013; Xie YB et al. 2014
Co-seismic variation: infrasonic radiation	Xu Q et al. 2013
Co-seismic variation: high-rate GPS recordings	Lou YD et al. 2014
Disasters: surface rupture	Han ZJ et al. 2013; Li YS et al. 2013; Liu MJ et al. 2014
Disasters: strong ground motion and seismic intensity	Dai ZJ et al. 2013; Luo YH et al. 2013; Meng LY et al. 2013, 2014; Ren YF et al. 2013; Wan XH et al. 2013; Wang YS et al. 2013; Wen RZ et al. 2013a, b; Zhang DL et al. 2013; Zhu GS et al. 2013; Xie JJ et al. 2014
Disasters: engineering damage and site effects	Dai JW et al. 2013; Gao YW et al. 2013; Gong MS et al. 2013; Jin B et al. 2013; Li ZQ et al. 2013; Liu JL et al. 2013a, b; Liu RS et al. 2013; Qi WH et al. 2013a; Qi WH et al. 2013b; Qu Z et al. 2013; Sun BT et al. 2013; Wang T et al. 2013a, b; Wang YM et al. 2013; Xiong LH et al. 2013; Xu C et al. 2013a; Ye F et al. 2013a, b; Zhang LX et al. 2013; Zhou TG et al. 2013
Disasters: observation and monitoring, and disaster impact assessment	Chen K et al. 2013; Li CS et al. 2013; Li ZF et al. 2013; Xu ZH et al. 2013; Meng QX and Xu WY 2014; Tao ZR, 2014
Disasters: evacuation	Ma TF et al. 2013
Disasters: secondary disasters, especially rock bursts and landslides	Chang M et al. 2013; Chen XL et al. 2013; Chen XQ et al. 2013; Feng C, 2013; Hong HC et al. 2013; Lan HX et al. 2013; Li WL et al. 2013; Li X et al. 2013; Liu R et al. 2013; Pei XJ and Huang RQ 2013; Shen J et al. 2013; Song MQ et al. 2013; Wu WW et al. 2013; Xu C, 2013; Xu C and Xiao JZ 2013; Xu C et al. 2013b; Yang ZY, 2013; Zhang YH et al. 2013

(continued)

Table A1.1 (continued)

Topic	Authors
Aftershock sequence: relocation	Chen C and Xu Y 2013; Fang LH et al. 2013; Su JR et al. 2013; Zhang GW and Lei JS 2013; Sun Z et al. 2014
Aftershock sequence: focal mechanism	Lü J et al. 2013; Zhao B et al. 2013; Han LB et al. 2014
Aftershock sequence: stress triggering	Miao M and Zhu SB 2013
Aftershock sequence: ETAS analysis and decay rate	Jiang CS et al. 2013; Jia Z et al. 2013
Field investigation and study: tectonics and stress state	Chen LC et al, 2013; Gong Y et al. 2013; Li CY et al. 2013; Li YS et al. 2013; Liu S et al. 2013; Xu XW et al. 2013b; Lei JS et al. 2014
Field investigation and study: simulation	Li YJ et al. 2013; Zhang ZQ et al. 2013a, b
Field investigation and study: deep structure and fault zone property	Chen S et al. 2013; Li ZW et al, 2013; Shen XZ, 2013; Sun Y and Lai XL 2013; Tian BF et al. 2013; Wen J and Chen XF 2013; Zheng Y et al. 2013
Field investigation and study: the reservoir triggering debate	Cheng HH et al. 2013
Relation with the Wenchuan earthquake: seismology and geology	Chen YT et al. 2013a, b; Du F, 2013; Xu XW et al. 2013a; Ying DL et al. 2013; Zhang YQ et al. 2013; Chen L et al. 2014
Relation with the Wenchuan earthquake: stress interaction	Lei XL et al. 2013; Liu BY et al. 2013; Wu JC et al. 2013; Parsons T and Segou M 2014; Wang Y et al. 2014
Relation with the Wenchuan earthquake: ETAS	Jia K et al. 2014
Seismic hazard: moment deficit	Liu M et al. 2014
Seismic hazard: deformation and locking	Jiang FY et al. 2013; Meng XG and Liu ZG 2013; Niu AF et al. 2013a; Niu AF et al. 2013b; Zhao J et al. 2013
Seismic hazard: pattern informatics	Jiang H et al. 2013.
Seismic hazard: CFS variation	Dong PY, 2013; Shan B et al. 2013; Zhu RH et al. 2013
Seismic hazard: the seismic gap	Gao Y et al. 2013
Precursor-like anomalies: seismicity and seismic waves	Rong DL and Li YR 2013; Wang LY et al. 2014
Precursor-like anomalies: gravity	Liang WF et al. 2013a; Liang WF et al. 2013b; Shi L et al. 2013; Zhu YQ et al. 2013; Zou ZB et al. 2013; Jiang L et al. 2014

(continued)

Table A1.1 (continued)

Topic	Authors
Precursor-like anomalies: OLR	Guo X et al. 2013
Precursor-like anomalies: electromagnetic	An ZH et al. 2013; Ma QZ et al. 2013; Xie T et al. 2013
Precursor-like anomalies: borehole strain	Chi SL, 2013; Chi SL et al. 2013; Qiu ZH et al. 2013
Precursor-like anomalies: infrasonic	Pan LL et al. 2013
Precursor-like anomalies: ionosphere	Jiang WP et al. 2013
Precursor-like anomalies: infrared thermal and ground temperature	Chen SY et al. 2013; Xu XD and Xu BH 2013; Zhang X et al. 2013a
Precursor-like anomalies: cloud	Wu LX et al. 2013
Precursor-like anomalies: comprehensive discussion and possible macro-anomalies	Guo AN, 2013; Li DW et al. 2013; Su XH et al. 2013; Zeng ZX and Wang J 2013; Zhang QY et al. 2013; Zhang X et al. 2013

See: http://www.cea-igp.ac.cn/lsdzzl/index.shtml; http://www.wceq.org/

In this table, the citation uses the surname plus the given-name-character abbreviation of the first author, so that different authors with the same surname (which is common in Chinese names) could be correctly differentiated. Note that majority of Chinese names are three characters, for example, the name Jiang (surname) Changsheng (first name) is abbreviated as Jiang CS. Additionally, in the reference list, for the journals which restart their page numbers issue by issue, the issue number is added to the volume number

References

An ZH, Du XB, Tan DC, Fan YY, Liu J, Cui TF (2013) Study on the geo-electric field variation of Sichuan Lushan M_S7.0 and Wenchuan M_S8.0 earthquake. Chin J Geophys 56:3868–3876. doi: 10.6038/cjg20131128 (in Chinese with English abstract)

Cai H, Zhao GQ (2013) Coseismic displacements and ionospheric changes of the 2013 MS7.0 Lushan earthquake from GPS measurements. Geodesy Geodynam 4(3):30–34. doi: 10.3724/SP.J.1246.2013.03030

Chang M, Tang C, Li WL, Zhang DD, Jia T, Ma GC, Zhou ZY (2013) Image interpretation and spatial analysis of geohazards induced by "4·20" Lushan earthquake in epicenter area. J Chengdu Univ Technol (Science & Technology Edition) 40:275–281. doi: 10.3969/j.issn.1671-9727.2013.03.07 (in Chinese with English abstract)

Chen C, Xu Y (2013) Relocation of the Lushan M_S7.0 earthquake sequence and its tectonic implication. Chin J Geophys 56:4028–4036. doi: 10.6038/cjg20131208 (in Chinese with English abstract)

Chen K, Yu YX, Gao MT, Gao JC (2013) Shakemap of peak ground acceleration with bias correction for the Lushan, Sichuan earthquake on April 20, 2013. Seismol Geol 35:627–633 (in Chinese with English abstract)

Chen L, Wang H, Ran Y, Lei S, Li X, Wu F, Ma X, Liu C, Han F (2014) The 2013 Lushan M_S7.0 earthquake: varied seismogenic structure from the 2008 Wenchuan earthquake. Seism Res Lett 85:34–39. doi: 10.1785/0220130109

Chen LC, Ran YK, Wang H, Li YB, Ma XQ (2013) The Lushan M_S7.0 earthquake and activity of the southern segment of the Longmenshan fault zone. Chin Sci Bull 58:3475–3482. doi: 10.1007/s11434-013-6009-6

Chen S, Xu WM, Shi L, Lu HY, Guo FY (2013) Gravity field and lithospheric mechanical properties of Longmenshan fault zone and its surrounding areas. Acta Seismol Sin 35:692–703. doi: 10.3969/j.issn.0253-3728.2013.05.008 (in Chinese with English abstract)

Chen SY, Liu PX, Liu LQ, Ma J (2013) A phenomenon of ground temperature change prior to Lushan earthquake observed in Kangding. Seismol Geol 35:634–640. doi: 10.3969/j.issn.0253-4967.2013.03.017 (in Chinese with English abstract)

Chen WW, Wang DC, Wei SJ (2013) A study on the uncertainties of the centroid depth of the 2013 Lushan earthquake from teleseimic body wave data. Earthq Sci 26(3–4):161–168. doi: 10.1007/s11589-013-0042-z

Chen XL, Yuan RM, Yu L (2013) Applying the Newmark's model to the assessment of earthquake-triggered landslides during the Lushan earthquake. Seismol Geol 35:661–670. doi: 10.3969/j.issn.0253-4967.2013.03.019 (in Chinese with English abstract)

Chen XQ, Cui P, You Y, Yang ZJ, Kong YD (2013) Secondary mountain disasters induced by the 4.20 Lushan earthquake and disaster mitigation. Earth Sci Front 20(3):29–34 (in Chinese with English abstract)

Chen YT, Yang ZX, Zhang Y, Liu C (2013a) A brief talk on the 20 April 2013 Lushan M_W6.7 earthquake. Acta Seismol Sin 35:285–295. doi: 10.3969/j.issn.0253-3782.2013.03.001 (in Chinese with English abstract)

Chen YT, Yang ZX, Zhang Y, Liu C (2013b) From 2008 Wenchuan earthquake to 2013 Lushan earthquake. Scientia Sinica Terrae 43:1064–1072 (in Chinese)

Cheng HH, Zhang H, Shi YL (2013) Did the occurrence of M_S7.0 Lushan earthquake relate to reservoir? Sci Technol Rev 31(31):31–37. doi: 10.3981/j.issn.1000-7857.2013.31.004 (in Chinese with English abstract)

Chi SL (2013) Strain anomalies before Wenchuan and Lushan earthquakes recorded by component before strainmeter. Sci Technol Rev 31(12):27–30. doi: 10.3981./j.issn.1000-7857.2013.12.004 (in Chinese with English abstract)

Chi SL, Liu Q, Chi Y, Deng T, Liao CW, Yang G, Zhang GP, Chen J (2013) Borehole strain anomalies before the 20 April 2013 Lushan MS7.0 earthquake. Acta Seismol Sin 35:296–303. doi: 10.3969/j.issn.0253-3782.2013.03.002 (in Chinese with English abstract)

Dai JW, Qu Z, Zhang CX, Weng XR (2013) Preliminary investigation of seismic damage to two steel space structure during the 2013 Lushan earthquake. Earthq Eng Eng Vibrat 12(3):497–500. doi: 10.1007/s11803-013-0189-6

Dai ZJ, Li XJ (2013) An explanation of the large PGA value of 2013 M_S7.0 Lushan earthquake at 51BXD station through topographic analysis. Earthq Sci 26(3–4): 199–205. doi: 10.1007/s11589-013-0043-y

Dong PY, Cheng HH, Zeng XF, Shi YL (2013) Stress changes on major faults by M_S7.0 Lushan earthquake. Sci Technol Rev 31(12):19–22. doi: 10.3981/j.issn.1000-7857.2013.12.002 (in Chinese with English abstract)

Du F, Long F, Ruan X, Yi GX, Gong Y, Zhao M, Zhang ZW, Qiao HZ, Wang Z, Wu J (2013) The M7.0 earthquake and the relationship with the M8.0 Wenchuan earthquake in Sichuan, China. Chin J Geophys 56:1772–1783. doi: 10.6038/cjg20130535 (in Chinese with English abstract)

Du YJ, Wang ZM, Yang SJ, An JC, Liu Q, Che GW (2013) Co-seismic deformation derived from GPS observations during April 20th, 2013 Lushan Earthquake, Sichuan, China. Earthq Sci 26:153–160. doi: 10.1007/s11589-013-0014-3

Fang LH, Wu JP, Wang WL, Lü ZY, Wang CZ, Yang T, Cai Y (2013) Relocation of mainshock and aftershock sequences of M_S7.0 Sichuan Lushan earthquake. Chin Sci Bull 58:3451–3459. doi: 10.1007/s11434-013-6000-2

Feng C, Liu R, Gou CJ (2013) Research of landslide susceptibility after the Lushan earthquake based on Logistic regression model. J Chengdu Univ Technol (Science & Technology Edition) 40:282–287. doi: 10.3969/j.issn.1671-9727.2013.03.08 (in Chinese with English abstract)

Gao Y, Wang Q, Zhao B, Shi YT (2013) A rupture blank zone in middle south part of Longmenshan faults: effect after Lushan M_S7.0 earthquake of 20 April 2013 in Sichuan, China. Science China: Earth Sci 43:1038–1046. doi: 10.1007/s11430-013-4646-x (in Chinese)

Gao YW, Jin B, Dai JW, Chen HF (2013) Damage to rural wooden structures in M7.0 Lushan earthquake and reinforcement suggestions. J Earthq Eng Eng Vibrat 33(3):48–52. doi: 10.11810/100-1301.20130306 (in Chinese with English abstract)

Gong MS, Yang YQ, Xie LL (2013) Seismic damage to reinforced concrete frame buildings in Lushan M7.0 earthquake. J Earthq Eng Eng Vibrat 33(3):20–26. doi: 10.11810/1000-1301.20130303 (in Chinese with English abstract)

Gong Y, Long F, Yi GX, Zhao M, Qiao HZ, Wang YX (2013) Study of apparent stress on the southern segment of Longmenshan fault zone around the Lushan earthquake. Technol Earthq Disaster Prevent 8:389–396 (in Chinese with English abstract)

Guo AN, Li X, Zhao CC, Zhang WC (2013) Retrospective Study on one-year prediction for 2013 Lushan M_S7.0 Earthquake. China Earthq Eng J 2:257–265. doi: 10.3969/j.issn.1000-0844.2013.02.0257 (in Chinese with English abstract)

Guo X, Zhang YS, Wei CX, Zhong MJ, Zhang X (2013) OLR anomalies for the Lushan MS7.0 earthquake. Acta Seismol Sin 35:731–737. doi: 10.3969/j.issn.0253-3782.2013.05.011 (in Chinese with English abstract)

Han LB, Zeng XF, Jiang CS, Ni SD, Zhang HJ, Long F (2014) Focal mechanisms of the 2013 M_W6.6 Lushan, China earthquake and high-resolution aftershock relocations. Seism Res Lett 85:8–14. doi: 10.1785/0220130083

Han ZJ, Ren ZK, Wang H, Wang MM (2013) The surface rupture signs of the Lushan "4·20" M_S7.0 earthquake at Longmen township, Lushan county and its discussion. Seismol Geol 35:388–397. doi: 10.3969/j.issn.0253-4967.2013.02.017 (in Chinese with English abstract)

Hao JL, Ji C, Wang WM, Yao ZX (2013) Rupture history of the 2013 M_W 6.6 Lushan earthquake constrained with local strong motion and teleseismic body and surface waves. Geophys Res Lett 40:5371–5376. doi: 10.1002/2013GL056876, 2013

Hu XG, Jiang Y (2013) Evaluation of the seismic moment of the April 20, 2013 Lushan earthquake. Earthq Sci 26:169–177. doi: 10.1007/s11589-013-0029-9

Hong HC, Xu HG, Song F, Zhang XH, Li GH (2013) Discussion on seismo-geological hazards incluced by 2013 Lushan MS7.0 earthquake and its seismogenic fault. Acta Seismol Sin 35:738–748. doi: 10.3969/j.issn.0253-3782.2013.05.012 (in Chinese with English abstract)

Jia K, Zhou SY, Zhuang JC, Jiang CS (2014) Possibility of the independence between the 2013 Lushan earthquake and the 2008 Wenchuan earthquake on Longmen Shan fault, Sichuan. Seism Res Lett 85:60–67. doi: 10.1785/0220130115

Jia Z, Chen WW, Chu RS (2013) Preliminary study on aftershock decay rate of the 2013 M_S7.0 Lushan earthquake. Earthq Sci 26:185–190. doi: 10.1007/s11589-013-0050-z

Jiang CS, Zhuang JC, Long F, Han LB, Guo LJ (2013) Statistical analysis of ETAS parameters in the early stage of the 2013 Lushan M_S7.0 earthquake sequence. Acta Seismol Sin 35:661–669. doi: 10.3969/j.jssn.0253-3782.2013.05.005 (in Chinese with English abstract)

Jiang FY, Zhu LY, Zhang XL, Wang SX, Ji LY (2013) Deformation preparation background of Lushan M_S7.0 earthquake and its coseismal response. J Seismol Res 36:450–454 (in Chinese with English abstract)

Jiang H, Wu ZL, Ma TF, Jiang CS (2013) Retrospective test of the PI forecast: case study of the April 20, 2013, Lushan, Sichuan, China, M_S7.0 earthquake. Physics 42:334–340. doi: 10.7693/wl20130504 (in Chinese with English abstract)

Jiang L, Li DQ, Xu ZP, Yang LP, Xiong W, Xu SQ (2014) Characteristics of gravity field before the Lushan M_S7.0 earthquake based on time variable gravity data from GRACE satellite. Acta Seismol Sin 36:84–94. doi: 10.3969/j.issn.0253-3782.2014.01.007

Jiang WP, Ma YF, Liu HF, Deng LS, Zhou XH (2013) Investigation of Lushan earthquake ionosphere VTEC anomalies based on GPS data. Earthq Sci 26:259–265. doi: 10.1007/s11589-013-0013-4

Jin B, Gao YW, Chen HF, Liu GQ, Wang T (2013) Primary damage analysis and study of the"4.20" Lushan earthquake on single-story factory buildings with brick column. J Nat Disasters 22(3):1–8 (in Chinese with English abstract)

Jin MP, Wang RJ, Tu HW (2014) Slip model and co-seismic displacement field derived from near-source strong motion records of the Lushan M_S7.0 earthquake on 20 April 2013. Chin J Geophys 57:129–137. doi: 10.6038/cjg20140102 (in Chinese with English abstract)

Lan HX, Zhou CH, Gao X, Cheng WM, Wang ZH, Yang ZH, Li LP, Wu YM (2013) Secondary geological hazard assessment and hazard mitigation countermeasures in Lushan, Ya'an Earthquake, Sichuan Province. Prog Geophys 32:499–504. doi: 10.11820/dlkxjz.2013.04.001

Lei JS, Zhang GW, Xie FR (2014) The 20 April 2013 Lushan, Sichuan, mainshock, and its aftershock sequence: tectonic implications. Earthq Sci 27:15–25. doi: 10.1007/s11589-0045-9

Lei XL, Ma SL, Su JR, Wang XL (2013) Inelastic triggering of the 2013 M_W6.6 Lushan earthquake by the 2008 M_W7.9 Wenchuan earthquake. Seismol Geol 35:411–422. doi: 10.3969/j.issn.0253-4967.2013.02.019 (in Chinese with English abstract)

Li CS, Yang JS, Tian BF, Jiang XD, Xu ZQ (2013) Rapid assessment of direct economic loss caused by M_S7.0 Lushan, Sichuan earthquake. J Nat Disasters 22(3):9–17 (in Chinese with English abstract)

Li CY, Xu XW, Gan WJ, Wen XZ, Zheng WJ, Wei ZY, Xu C, Tan XB, Chen GH, Liang MJ, Li XN (2013) Seismogenic structures associated with the 20 April 2013 M_S7.0 Lushan earthquake, Sichuan Province. Seismol Geol 35:671–683. doi: 10.3969/j.issn.0253-4967.2013.03.020 (in Chinese with English abstract)

Li DW, Chen GF, Chen JL, Hao HJ (2013) Earthquake prediction: from Lushan earthquake to continental earthquakes. Earth Sci Frontiers 20(3):1–10. (in Chinese with English abstract)

Li X, Liu XL, Li JG, Wang QL, Liao WL, Zhang LF (2013) Factor analysis of earthquake-induced geological disasters of the M7.0 Lushan earthquake in China. Geodesy Geodynam 4(2):22–29. doi: 10.3724 /SP.J.1246.2013.02022

Li WL, Huang RQ, Xu Q, Tang C (2013) Rapid prediction of co-seismic landslides triggered by Lushan earthquake, Sichuan, China. J Chengdu Univ Technol (Science & Technology Edition) 40:264–274. doi: 10.3969/j.issn.1671-9727.2013.03.06 (in Chinese with English abstract)

Li YJ, Chen LW, Yang SX (2013) Numerical simulation of the co-seismic effect deduced from the Lushan strong earthquake: insights from the strain energy change. Technol Earthq Disaster Prevent 8:361–369 (in Chinese with English abstract)

Li YS, Wang YS, Pei XJ, Luo YH (2013) Research on tectonic fracturing and causative fault of Lushan earthquake in Sichuan, China. J Chengdu Univ Technol (Science & Technology Edition) 40:242–249. doi: 10.3969/j.issn.1671-9727.2013.03.03 (in Chinese with English abstract)

Li ZF, Wu LX, Wang R, Wang QL, Zheng S, Xu ZH, Li J (2013) Rapid assessment of earthquake disaster and rethink of M_S7.0 Lushan earthquake. Sci Technol Rev 31(12):31–36. doi: 10.3981/j.issn.1000-7857.2013.12.005 (in Chinese with English abstract)

Li ZQ, Hou JS, Li Y, Su GW, Jiang LX, Sun BT, Yuan YF, Lin JQ (2013) Analysis on the characteristics of the M_S7.0 Lushan, Sichuan Province, earthquake hazard on April 20, 2013. Seismol Geol 35:398–410. doi: 10.3969/j.issn.0253-4967.2013.02.018 (in Chinese with English abstract)

Li ZW, Tian BF, Liu S, Yang JS (2013) Asperity of the 2013 Lushan earthquake in the eastern margin of Tibetan Plateau from seismic tomography and aftershock relocation. Geophys J Int 195:2016–2022. doi: 10.1093/gji/ggt370

Liang WF, Liu F, Xu YM, Zhu YQ, Guo SS, Zhao YF, Liu L (2013a) Gravity observation in the eastern margin of Qinghai-Tibetan plateau and reflect to the Lushan M7.0 earthquake. China Earthq Eng J 2:266–271. doi: 10.3969/j.issn.1000-0844.2013.02.0266 (in Chinese with English abstract)

Liang WF, Zhao YF, Xu YM, Zhu YQ, Guo SS, Liu F, Liu L (2013b) Gravity observations along the eastern margin of the Tibetan plateau and an application to the Lushan M_S7.0 earthquake. Earthq Sci 26:251–257. doi: 10.1007/s11589-013-0024-1

Lin XD, Ge HK, Xu P, Dreger DS, Su JR, Wang BS, Wu MJ (2013) Near field full waveform inversion: Lushan magnitude 7.0 earthquake and its aftershocks moment tensor. Chin J Geophys 56:4037–4047. doi: 10.6038/cjg20131209 (in Chinese with English abstract)

Liu BY, Shi BP, Lei JS (2013) Effect of Wenchuan earthquake on probabilities of occurrence of Lushan and surrounding faults. Acta Seismol Sin 35:642–651. doi: 10.3969/j.issn.0253-3782.2013.05.003 (in Chinese with English abstract)

Liu CL, Zheng Y, Ge C, Xiong X, Xu HZ (2013) Rupture process of the M7.0 Lushan earthquake, 2013. Sci China: Earth Sci 56:1187–1192. doi: 10.1007/s11430-013-4639-9

Liu G, Zhao B, Zhang R, Huang Y, Wang J, Nie ZS, Qiao XJ, Tan K (2013) Near- field surface deformation during the April 20, 2013, M_S7.0 Lushan earthquake measured by 1- Hz GNSS. Geodesy Geodynam 4(2):1–5. doi: 10.3724/SP.J.1246.2013.02001

Liu J, Yi GX, Zhang ZW, Guan ZJ, Ruan X, Long F, Du F (2013) Introduction to the Lushan, Sichuan M7.0 earthquake on 20 April 2013. Chin J Geophys 56:1404–1407. doi: 10.6038/cig20130434 (in Chinese with English abstract)

Liu JL, Lin JQ, Liu RS, Guo ED, Ye F, Yan DQ (2013a) Investigation and analysis of seismic damage of highway system in Lushan "4.20" earthquake. J Nat Disasters 22(3):18–23 (in Chinese with English abstract)

Liu JL, Lin JQ, Liu RS, Guo ED, Ye F, Yan DQ (2013b) Seismic damage investigation and analysis of the communication system in Lushan M_S7.0 earthquake. J Nat Disasters 22(5):91–97 (in Chinese with English abstract)

Liu M, Luo G, Wang H (2014) The 2013 Lushan earthquake in China tests hazard assessments. Seism Res Lett 85:40–43. doi: 10.1785/0220130117

Liu MJ, Wang FY, Jia SX, Han YJ, Li XM, Feng JL, Sun H, Shen HL, Zhang JS, Tao H, Dai AP, Sun Y, Wang C (2014) Jianganlin surface fracture induced by 2013 Lushan M_S7.0 earthquake. Acta Seismol Sin 36:129–138. doi: 10.3969/j.issn.0253-3782.2014.01.011 (in Chinese with English abstract)

Liu R, Su W, Zhang XD, Tian LY, Zhang QY (2013) Hazard assessment of the secondary landslides after Lushan earthquake. Technol Earthq Disaster Prevent 8:423–433 (in Chinese with English abstract)

Liu RF, Chen YT, Zou LY, Chen HF, Liang JH, Zhang LW, Han XJ, Ren X, Sun L (2013) Determination of parameters for the 20 April 2013 Lushan M_W6.7(M_S7.0) earthquake in Sichuan Province. Acta Seismol Sin 35:652–660. doi: 10.3969/j.issn.0253-3782.2013.05.004 (in Chinese with English abstract)

Liu RS, Liu JL, Yan DQ, Ye F, Lin JQ, Li J (2013) Seismic damage investigation and analysis of electric power system in Lushan M_S7.0 earthquake. J Nat Disasters 22(5):83–90 (in Chinese with English abstract)

Liu S, Yang JS, Tian BF, Zheng Y, Jiang XD, Xu ZQ (2013) The characteristics of shear wave splitting in the source region of the April 20, 2013 Lushan earthquake. Earthq Sci 26:223–228. doi: 10.1007/s11589-013-0034-z

Lou YD, Zhang WX, Shi C, Liu JN (2014) High-rate (1-Hz and 50-Hz) GPS seismology: application to the 2013 M_W6.6 Lushan earthquake. J Asian Earth Sci 79:426–431

Luo YH, Wang YS, He Y, Gao Y, Liu Z, Cao WZ (2013) Monitoring result analysis of Lengzhuguan slope ground shock response of Lushan earthquake of Sichuan, China. J Chengdu Univ Technol (Science & Technology Edition) 40:232–241. doi: 10.3969/j.issn.1671-9727.2013.03.02 (in Chinese with English abstract)

Lü J, Wang XS, Su JR, Pan LS, Li Z, Yin LW, Zeng XF, Deng H (2013) Hypocentral location and source mechanism of M_S7.0 Lushan earthquake sequence. Chin J Geophys 56:1753–1763. doi: 10.6038/cjg20130533 (in Chinese with English abstract)

Ma QZ, Fang GQ, Li W, Zhou JN (2013) Electromagnetic anomalies before the 2013 Lushan M_S7.0 earthquake. Acta Seismol Sin 35:717–730. doi: 10.3969/j.issn.0253-3782.2013.05.010 (in Chinese with English abstract)

Ma TF, Wu ZL, Yang XL (2013) Video recording of classroom evacuation during the April 20, 2013, Lushan, Sichuan, China, earthquake: preliminary analysis and implication for the statistical physical models of escape panic. Physics 42:341–345. doi: 10.7693/wl20130505 (in Chinese with English abstract)

Meng LY, Zhou LQ, Liu J (2013) Source parameters of the 2013 Lushan M_S7.0 earthquake and the characteristics of the near-fault strong ground motion. Acta Seismol Sin 35:632–641. doi: 10.3969/j.issn.0253-3782.2013.05.002 (in Chinese with English abstract)

Meng LY, Zhou LQ, Liu J (2014) Estimation of the near-fault strong ground motion and intensity distribution of the 2013 Lushan, Sichuan, M_S7.0 earthquake. Chin J Geophys 57:441–448. doi: 10.6.38/cjg20140210 (in Chinese with English abstract)

Meng QX, Xu WY (2014) Ya'an earthquake of 20 April 2013: introduction and reflections. Nat Hazards 70:941–949. doi: 10.1007/s11069-013-0817-5

Meng XG, Liu ZG (2013) The M_S7.0 Lushan earthquake and the activity of the Longmenshan fault zone. Geodesy Geodynam 4(3):40–47. doi: 10.3724/SP.J1246.2013.03040

Miao M, Zhu SB (2013) The static Coulomb stress change of the 2013 Lushan M_S7.0 earthquake and its impact on the spatial distribution of aftershocks. Acta Seismol Sin 35:619–631. doi: 10.3969/j.issn.0253-3782.2013.05.001 (in Chinese with English abstract)

Niu AF, Gu GH, Cao JP, Zhang LK, Yan W, Zhao J, Ji P (2013a) On the preseismic deformation changes prior to the Lushan M_S7.0 earthquake. Acta Seismol Sin 35:670–680. doi: 10.3969/j.issn.0253-3782.2013.05.006 (in Chinese with English abstract)

Niu AF, Zhang LK, Li Y, Yan W, Zhao J, Ji P (2013b) On the weak deformation area prior to the Lushan M_S7.0 earthquake. Technol Earthq Disaster Prevent 8:341–350 (in Chinese with English abstract)

Pan LL, Zeng ZX, Wang J (2013) Precursory abnormal infrasonic signals predating the Lushan earthquake (M_S7.0) and Yushu earthquake (M_S5.2). Earth Sci Frontiers 20(6):73–79 (in Chinese with English abstract)

Parsons T, Segou M (2014) Stress, distance, magnitude, and clustering influences on the success or failure of an aftershock forecast: the 2013 M=6.6 Lushan earthquake and other examples. Seism Res Lett 85:44–51. doi: 10.1785/0220130100

Pei XJ, Huang RQ (2013) Analysis of characteristics of geological hazards by "4·20" Lushan earthquake in Sichuan, China. J Chengdu Univ Technol (Science & Technology Edition) 40:257–263. doi: 10.3969/j.issn.1671-9727.2013.03.05 (in Chinese with English abstract)

Qi WH, Chen LW, Shan ZD, Wang YL (2013a) Summary of macro-level site effects in M_S7.0 Lushan earthquake. J Earthq Eng Eng Vibrat 33(4):29–34. doi: 10.11810/1000-1301.20130404 (in Chinese with English abstract)

Qi WH, Chen LW, Wen Z, Shan ZD, Wang YL (2013b) Preliminary discussion on ground and foundation damages in *M*s7.0 Lushan earthquake. J Nat Disasters 22(5):98–103 (in Chinese with English abstract)

Qiu ZH, Yang G, Tang L, Guo YP, Zhang BH (2013) Abnormal strain changes observed by a borehole strainmeter at Guza Station before the M_S7.0 Lushan earthquake. Geodesy Geodynam 4(3):19–29. doi: 10.3724/SP.J.1246.2013.03019

Qu Z, Zhong JR, Sun JJ (2013) Seismic damage to masonry structures in *M*7.0 Lushan earthquake. J Earthq Eng Eng Vibrat 33(2):27–35. doi: 10.11810/1000-1301.20130304 (in Chinese with English abstract)

Ren YF, Ji K, Wen RZ (2013) Field survey around typical strong motion stations and its explanation on the macro seismic intensity in Lushan earthquake on April 20, 2013. Earthq Sci 26:241–250. doi: 10.1007/s11589-013-0051-y

Rong DL, Li YR (2013) The critical point behavior of seismic activities prior to Lushan M_S7.0 earthquake in 2013. China Earthq Eng J 2:252–256. doi: 10.3969/j.issn.1000-0844.2013.02.0252 (in Chinese with English abstract)

Shan B, Xiong X, Zheng Y, Jin BK, Liu CL, Xie ZJ, Xu HZ (2013) Stress changes on major faults caused by 2013 Lushan earthquake, and its relationship with 2008 Wenchuan earthquake. Sci China Earth Sci 56:1169–1176. doi: 10.1007/s11430-013-4642-1

Shen J, Bo JS, Yu XH, Lu T (2013) Seismotectonics and seismogeological disaster of the April 20, 2013 M_S7.0 (M_W6.6) Lushan earthquake in Sichuan Province of China. J Inst Disaster-Prevent Sci Technol 15(3):1–8 (in Chinese with English abstract)

Shen XZ (2013) An analysis of the deformation of the crust and LAB beneath the Lushan and Wenchuan earthquakes in Sichuan Province. Chin J Geophys 56:1895–1903. doi: 10.6038/cjg20130612 (in Chinese with English abstract)

Shi L, Chen S, Jiang CS, Xu WM, Lu HY, Guo FY (2013) Characteristics of gravity anomaly in Lushan earthquake zone based on preferential filtering method. Acta Seismol Sin 35:704–716. doi: 10.3969/j.issn.0253-3782.2013.05.009 (in Chinese with English abstract)

Song MQ, Zheng Y, Fu R, Liang XJ, Wang X, Li L, Li B (2013) Did the M_S7.0 Lushan earthquake dynamically trigger earthquakes in the Datong volcanic region (Shanxi Province)? Earthq Sci 26:229–239. doi: 10.1007/s11589-013-0035-y

Su JR, Zheng Y, Yang JS, Chen TC, Wu P (2013) Accurate location of the Lushan, Sichuan *M*7.0 earthquake on 20 April 2013 and its aftershocks and analysis of the seismogenic structure. Chin J Geophys 56:2636–2644. doi: 10.6038/cjg20130831 (in Chinese with English abstract)

Su XH, Zhang QY, Zhang XD, Zhang X Li L, Liu JM, Huang JX, Su W (2013) Analysis on macroscopic anomalies before and after Lushan M_S7.0 earthquake based on weibo. Technol Earthq Disaster Prevent 8:451–458 (in Chinese with English abstract)

Sun BT, Wang MZ, Yan PL, Zhang HY, Chen XZ (2013) Damage characteristics and seismic analysis of single-story brick bent frame column industrial buildings in Lushan M_S7.0 earthquake. J Earthq Eng Eng Vibrat 33(3):1–8. doi: 10.11810/1000-1301.20130301 (in Chinese with English abstract)

Sun XD, Wang G, Liu CQ (2013) Finite fault models of the Lushan earthquake made by using hybrid slip model. J Earthq Eng Eng Vibrat 33(4):15–20. doi: 10.11810/1000-1301.20130402 (in Chinese with English abstract)

Sun Y, Lai XL (2013) Fault zone trapped waves at Longmenshan fault belt. Geodesy Geodynam 4(3):48–52. doi: 10.3724/SP.J.1246.2013.03048

Sun Z, Wu JP, Fang LH, Wang WL, Wang CZ, Yang T (2014) Focal depth determination of aftershocks of Lushan *M*s7.0 earthquake from sPn phase. Chin J Geophys 57:430–440. doi: 10.6038/cjg20140209 (in Chinese with English abstract)

Tan HB, Shen CY, Wei J, Zhao B, Wang J, Xuan SB (2013) Simulation of coseismic effects of the M_S7.0 Lushan earthquake. Geodesy Geodynam 4(3):12–18. doi: 10.3724/SP.J.1246.2013.03012

Tang L, Jing Y (2013) Analysis of coseismic strain step observed by 4-component borehole strain meters in Sichuan-Yunnan region. Technol Earthq Disaster Prevent 8:370–376 (in Chinese with English abstract)

Tao ZR (2014) Short-term economic effect of the *M*7.0 Lushan earthquake. Nat Hazards 70:1247–1261. doi: 10.1007/s11069-013-0871-z

Tian BF, Li ZW, Liu S, Yang JS (2013) High-resolution seismic tomography of the upper crust in the source region of the April 20, 2013 Lushan earthquake and its seismotectonic implications. Earthq Sci 26:213–221. doi: 10.1007/s11589-013-0012-5

Wan XH, Shi YC, Lu YX, Liu K, Sheng ZQ, Li R, Tu HW (2013) Research on the dynamic change of acceleration and seismic intensity distribution of the Lushan M_S7.0 earthquake. China Earthq Eng J 35:900–907 (in Chinese with English abstract)

Wang K, Liu CL, Xiong X, Zheng Y (2013) Co- and post-seismic surface deformation and gravity changes of M_S7.0 Lushan, earthquake. Earthq Sci 26:207–212. doi: 10.1007/s11589-013-0001-8

Wang LY, Li YE, Zheng XY, Wang SW (2014) Temporal variation of v_p/v_s at single seismic station before the 2013 Lushan M_S7.0 earthquake. Acta Seismol Sin 36:42–58. doi: 10.3969/j.issn.0253-3782.2014.01.004 (in Chinese with English abstract)

Wang T, Lei YD, Zhang YQ, Kong ZA (2013a) Seismic damage analysis of buildings in Lushan earthquake. J Earthq Eng Eng Vibrat 33(2):36–47. doi: 10.118101/1000-1301.20130305 (in Chinese with English abstract)

Wang T, Zhang YQ, Jin B, Kong ZA, Lei YD (2013b) Seismic damage to masonry residential buildings in Lushan M_S7.0 earthquake. J Earthq Eng Eng Vibrat 33(3):9–19. doi: 10.11810/1000-1301.20130302 (in Chinese with English abstract)

Wang WM, Hao JL, Yao ZX (2013) Preliminary result for rupture process of Apr. 20, 2013, Lushan Earthquake, Sichuan, China. Chin J Geophys 56:1412–1417. doi: 10.6038/cjg20130436 (in Chinese with English abstract)

Wang Y, Wang F, Wang M, Shen Z-K, Wan Y (2014) Coulomb stress change and evolution induced by the 2008 Wenchuan earthquake and its delayed triggering of the 2013 M_W6.6 Lushan earthquake. Seism Res Lett 85:52–59. doi: 10.1785/0220130111

Wang YM, Xing LH, Xu WX (2013) Seismic damage and damage enlightenment of medical buildings in Lushan M_S7.0 earthquake. J Earthq Eng Eng Vibrat 33(4):44–53. doi: 10.11810/1000-1301.20130406 (in Chinese with English abstract)

Wang YS, Li XJ, Mei ZH, Liu Y (2013) Reliability comparison of instrumental seismic intensity algorithms in 2008 Wenchuan earthquake and 2013 Lushan earthquake. Acta Seismol Sin 35:759–770. doi: 10.3969/j.issn.0253-3782.2013.05.014 (in Chinese with English abstract)

Wei J, Zhao B, Tan HB, Yu D, Shen CY, Li H (2013) Detection of a half-microgal coseismic gravity change after the M_S7.0 Lushan earthquake. Geodesy Geodynam 4(3):7–11. doi: 10.3724/SP.J.1246.2013.03007

Wen J, Chen XF (2013) f_{max} and fault zone property of Lushan earthquake of 20 April 2013, Sichuan, China. Earthq Sci 26:179–183. doi: 10.1007/s11589-013-0017-0

Wen RZ, Ren YF, Huang XT, Lu T, Qi WH (2013a) Strong motion records and their engineering damage implications for Lushan earthquake on April 20, 2013. J Earthq Eng Eng Vibrat 33(4):1–14. doi: 10.11810/1000-1301.20130401 (in Chinese with English abstract)

Wen RZ, Ren YF, Qi WH, Lu T, Yang ZY, Shan ZD, Wang YL (2013b) Maximum acceleration recording from Lushan earthquake on April 20, 2013. J Southwest Jiaotong Univ 48(5):783–791. doi: 10.3969/j.issn.0258-2724.2013.05.001 (in Chinese with English abstract)

Wu JC, Yu S, Cai YJ, Lei DN, Li H (2013) Stress triggering of the Lushan *M*7.0 earthquake by the Wenchuan M_S8.0 earthquake. Geodesy Geodynam 4(3):35–39. doi: 10.3724/SP.J.1246.2013.03035

Wu LX, Zheng S, Zhong XH, Qin K (2013) Satellite linear clouds anomaly before M_S7.0 Lushan earthquake. Sci Technol Rev 31(12):23–26. doi: 10.3981/j.issn.1000-7857.2013.12.003 (in Chinese with English abstract)

Wu WW (2013) Characteristics and countermeasures of geological hazards induced in Wenchuan and Lushan earthquakes, Sichuan Province. Technol Earthq Disaster Prevent 8:434–439 (in Chinese with English abstract)

Wu YQ, Jiang ZS, Wang M, Che S, Liao H, Li Q, Li, P, Yang YL, Xiang HP, Shao ZG, Wang WX, Wei WX, Liu XX (2013) Preliminary results of the co-seismic displacement and pre-seismic strain accumulation of the Lushan M_S7.0 earthquake reflected by the GPS surveying. Chin Sci Bull 58:1910–1916. doi: 10.1007/s11434-013-5998-5

Xie JJ, Li XJ, Wen ZP, Wu CQ (2014) Characteristics of near-source vertical and horizontal strong ground motion from the 20 April 2013 M_W6.8 Lushan earthquake in China. Seism Res Lett 85:23–33. doi: 10.1785/0220130121

Xie T, Lu J, Ren YX, Zhao MY (2013) Analysis on apparent resistivity variations of Ganzi station before 2013 Lushan M_S7.0 Earthquake. Technol Earthq Disaster Prevent 8:377–388 (in Chinese with English abstract)

Xie YB, Wu JC, Chen JP, Xu KK, Peng Y (2014) Perturbation analysis of the ionospheric TEC before and after the Lushan M_S7.0 earthquake. Acta Seismol Sin 36: 95-105. doi: 10.3969/j.issn.0253-3782.2014.01.008

Xie ZJ, Jin BK, Zheng Y, Ge C, Xiong X, Xiong C, Xu HZ (2013) Source parameters inversion of the 2013 Lushan earthquake by combining teleseismic waveforms and local seismograms. Sci China Earth Sci 56:1177–1186. doi: 10.1007/s11430-013-4640-3

Xiong LH, Lan RQ, Wang YM, Tian XM, Feng B (2013) Earthquake damage investigation of structures in 7.0 Lushan strong earthquake. J Earthq Eng Eng Vibrat 33(4):35–43. doi: 10.11810/1000-1301.20130405 (in Chinese with English abstract)

Xu C (2013) Quick landslide susceptibly evaluation and its validation for the 2013 M_S7.0 Lushan earthquake. Sci Technol Rev 31(28/29):15–23. doi: 10.3981/j.issn.1000-7857.2013.h2.001 (in Chinese with English abstract)

Xu C, Xiao JZ (2013) Spatial analysis of landslides triggered by the 2013 M_S7.0 Lushan earthquake: a case study of a typical rectangle area in the northeast of Taiping town. Seismol Geol 35:436–451. doi: 10.3969/j.issn.0253-4967.2013.02.021 (in Chinese with English abstract)

Xu C, Chen B, Li XJ, Chen XL, Lü HS, Geng F, Wen ZP (2013a) Seismic performance characteristics and damage of buildings in Lushan M_S7.0 earthquake. Acta Seismol Sin 35:749–758. doi: 10.3969/j.issn.0253-3782.2013.05.013 (in Chinese with English abstract)

Xu C, Xu XW, Zheng WJ, Wei ZY, Tan XB, Han ZJ, Li CY, Liang MJ, Li ZQ, Wang H, Wang MM, Ren JJ, Zhang SM, He ZT (2013b) Landslides triggered by the April 20, 2013 Lushan, Sichuan province M_S7.0 strong earthquake of China. Seismol Geol 35:641–660. doi: 10.3969/j.issn.0253-4967.2013.03.018 (in Chinese with English abstract)

Xu LS, Yan C, Zhang X, Li CL (2013) Where did the Lushan M_S7.0 earthquake occur in the world? Chin J Geophys 56:2982–2993. doi: 10.6038/cjg20130912 (in Chinese with English abstract)

Xu Q, Zhu X, Li WL, Deng ML, Zheng G, Li ZH (2013) Study on infrasound waveforms from Lushan earthquake on April 20, 2013, Sichuan, China. J Chengdu Univ Technol (Science & Technology Edition) 40:225–231. doi: 10.3969/j.issn.1671-9727.2013.03.01 (in Chinese with English abstract)

Xu XD, Xu BH (2013) The infrared warming abnormality analysis of Lushan earthquake. Earth Sci Frontiers 20(3):25–28 (in Chinese with English abstract)

Xu XW, Chen GH, Yu GH, Cheng J, Tan XB, Zhu AL, Wen XZ (2013a) Seismogenic structure of Lushan earthquake and its relationship with Wenchuan earthquake. Earth Science Frontiers 20(3):11–20 (in Chinese with English abstract)

Xu XW, Wen XZ, Han ZJ, Chen GH, Li CY, Zheng WJ, Zhang SM, Ren ZK, Xu C, Tan XB, Wei ZY, Wang MM, Ren JJ, He Z, Liang MJ (2013b) Lushan M_S7.0 earthquake: a blind reserve-fault earthquake. Chin Sci Bull 58:3437–3443. 10.1007/s11434-013-5999-4

Xu Y, Shao WL (2013) Rupture details of the 20 April 2013 Lushan 2013 M7.0 earthquake. Chin J Geophys 56:3396–3403. doi: 10.6038/cjg20131015 (in Chinese with English abstract)

Xu ZH, Lu CB, Wang P, Wang QL, Li FS, Wu LX (2013) Airborne union observation and disaster enhanced identification of M_S7.0 Lushan earthquake. Sci Technol Rev 31(12):37–41. doi: 10.3981/j.issn.1000-7857.2013.12.006 (in Chinese with English abstract)

Yang ZY (2013) Features of secondary mountain hazards triggered by the "4·20" Lushan earthquake. J Sichuan Univ (Engineering Science Edition) 45(4):76–83 (in Chinese with English abstract)

Ye F, Guo ED, Liu JL, Liu RS, Lin JQ, Gao L (2013a) Seismic damage investigation and analysis of water supply system in Lushan earthquake. World Earthq Eng 29(4):52–58 (in Chinese with English abstract)

Ye F, Guo ED, Liu JL, Liu RS, Lin JQ, Liu Z (2013b) Seismic damage investigation and analysis of gas system in Lushan M_S7.0 earthquake. J Nat Disasters 22(5):77–82 (in Chinese with English abstract)

Ying DL, Li ZQ, Zeng Q, Li ZJ, Wan SS, Hang WY, Li HK, Ying WF (2013) Preliminary analysis of causative faults of Lushan earthquake and Wenchuan earthquake in Sichuan, China. J Chengdu Univ Technol (Science & Technology Edition) 40:250–256. doi: 10.3969/j.issn.1671-9727.2013.03.04 (in Chinese with English abstract)

Zeng XF, Han LB, Shi YL (2013a) Focal mechanism solution of M_S7.0 Lushan earthquake, April 20, 2013. Sci Technol Rev 31(12):15–18. doi: 10.3981/j.issn.1000-7857.2013.12.001 (in Chinese with English abstract)

Zeng XF, Luo Y, Han LB, Shi YL (2013b) The Lushan M_S7.0 earthquake on 20 April 2013: a high-angle thrust event. Chin J Geophys 56:1418–1424. doi: 10.6038/cjg20130437 (in Chinese with English abstract)

Zeng ZX, Wang J (2013) Lushan earthquake: a successful case for medium-term prediction. Earth Sci Frontiers 20(3):21–24 (in Chinese with English abstract)

Zhang DL, Huang B, Zhang XB, Xu XW, Zheng WJ (2013) Strong ground motion distribution and simulation based on finite fault model of Lushan 7.0 earthquake on April 20, 2013. Seismol Geol 35:423–435. doi: 10.3969/j.issn.0253-4967.2013.02.020 (in Chinese with English abstract)

Zhang GW, Lei JS (2013) Relocations of Lushan, Sichuan strong earthquake (M_S7.0) and its aftershocks. Chin J Geophys 56:1764–1771. doi: 10.6038/cjg20130534 (in Chinese with English abstract)

Zhang LF, Iman F, Liao WJ, Li JG (2013) Source rupture process inversion of the 2013 Lushan earthquake, China. Geodesy Geodynam 4(2):16–21. doi: 10.3724/SP.J.1246.2013.02016

Zhang LX, Liu X, Liu JP (2013) Investigation and consideration of the staircase in Lushan earthquake. J Earthq Eng Eng Vibrat 33(4):21–28. doi: 10.11810/1000-1301.20130403 (in Chinese with English abstract)

Zhang QY, Huang JX, Zhang XD, Su XH, Zhang X (2013) Micro-blog application in extracting information of earthquake macro-anomalies. Technol Earthq Disaster Prevent 8:459–467 (in Chinese with English abstract)

Zhang X, Zhang XD, Huang JX, Su XH (2013) Field investigation and analysis of macro-anomalies before the Lushan M_S7.0 earthquake, Sichuan, China. Technol Earthq Disaster Prevent 8:440–450 (in Chinese with English abstract)

Zhang X, Zhang YS, Wei CX, Tian XF, Tang Q, Gao J (2013a) Analysis of thermal infrared anomaly before the Lushan M_S7.0 earthquake. China Earthq Eng J 35:272–277. doi: 10.3969/j.issn.1000-0844.2013.02.0272 (in Chinese with English abstract)

Zhang Y, Xu LS, Chen YT (2013) Rupture of the Lushan 4.20 earthquake and preliminary analysis on the disaster-causing mechanism. Chin J Geophys 56:1408–1411. doi: 10.6038/cjg20130435 (in Chinese with English abstract)

Zhang Y, Chen YT, Xu L, Du F, Jin M, Tu H, Dahm T (2014) Kinematic rupture model and hypocenter relocation of the 2013 M_W6.6 Lushan earthquake constrained by strong-motion and teleseismic data. Seism Res Lett 85:15–22. doi: 10.1785/0220130126

Zhang YH, Dong SW, Hou CT, Guo CB, Yao X, Li B, Du JJ, Zhang JG (2013) Geohazards induced by the Lushan M_S7.0 earthquake in Sichuan Province, southwest China: typical examples, types and distributional characteristics. Acta Geol Sin 87:646–657. doi: 10.1111/1755-6724.12076

Zhang YQ, Dong SW, Hou CT, Shi JS, Wu ZH, Li HL, Sun P, Liu G, Li J (2013) Seismogenic structure of the April 20, 2013, Lushan M_S7 earthquake in Sichuan. Acta Geol Sin 87:633–645

Zhang ZQ, Chen T, Ren ZK, Wang WT (2013a) Characters of superface displacement due to blind curved fault revealed by numerical models. Seismol Geol 35:452–460. doi: 10.3969/j.issn.0253-4967.2013.02.022

Zhang ZQ, Wang WT, Ren ZK, Zhang PZ, Fang LH, Wu JP (2013b) Lushan $M7.0$ earthquake: a special earthquake occurs on curved fault. Chin Sci Bull 58:3483–3490. doi: 10.1007/s11434-013-6010-0

Zhao B, Gao Y, Huang ZB, Jiang X, Li DH (2013) Double difference relocation, focal mechanism and stress inversion of Lushan $M_S7.0$ earthquake sequence. Chin J Geophys 56:3385–3395. doi: 10.6038/cjg20131014 (in Chinese with English abstract)

Zhao CP, Zhou LQ, Chen ZL (2013) Source rupture process of Lushan $M_S7.0$ earthquake, Sichuan, China and its tectonic implication. Chin Sci Bull 58:3444–3450. doi: 10.1007/s11434-013-6017-6

Zhao J, Wu YQ, Jiang ZS, Niu AF, Liu J, Wang LF, Wei WX (2013) Fault locking and dynamic deformation of the Longmenshan fault zone before the 2013 Lushan $M_S7.0$ earthquake. Acta Seismol Sin 35:681–691. doi: 10.3969/j.issn.0253-3782.2013.05.007 (in Chinese with English abstract)

Zhao X, Huang ZB, Fang LH, Li Q, Zhao B, Miao CL (2014) Kinematic characteristics of the source process of the Lushan, Sichuan $M_S7.0$ earthquake on 20 April 2013. Chin J Geophys 57:419–429. doi: 10.6038/cjg20140208 (in Chinese with English abstract)

Zheng Y, Ge C, Xie ZJ, Yang YJ, Xiong X, Xu HZ (2013) Crustal and upper mantle structure and the deep seismogenic environment in the source regions of the Lushan earthquake and the Wenchuan earthquake. Sci China Earth Sci 56:1158–1168. doi: 10.1007/s11430-013-4641-2

Zhou TG, Qian XB, Zhang BB (2013) Investigation and analysis of damage to rural houses in Lushan earthquake. J Earthq Eng Eng Vibrat 33:53–58. doi: 10.11810/1000-1301.20130307 (in Chinese with English abstract)

Zhou YY, Wu Y, Yang J, Zhu FY, Zhao B (2013) Ionospheric disturbances around the time of the $M_S7.0$ Lushan earthquake. Geodesy Geodynam 4(4):26–32. doi: 10.3724/SP.J.1246.2013.04026

Zhu GS, Zhang ZG, Wen J, Zhang W, Chen XF (2013) Preliminary results of strong ground motion simulation for the Lushan earthquake of 20 April 2013, China. Earthq Sci 26:191–197. doi: 10.1007/s11589-013-002-5

Zhu RH, Xie ZD, Zhao XY, Lei XL, Yan X (2013) Research on influence of Coulomb stress changes induced by Lushan $M_S7.0$ earthquake on eastern Yunnan areas. J Seismol Res 36:455–461 (in Chinese with English abstract)

Zhu YQ, Wen XZ, Sun HP, Guo SS, Zhao YF (2013) Gravity changes before the Lushan, Sichuan, M_S=7.0 Earthquake of 2013. Chin J Geophys 56:1887–1894. doi: 10.6038/cjg20130611 (in Chinese with English abstract)

Zou ZB, Li H, Kang KX, Wu YL (2013) Characteristics of satellite-gravity variations in the North-South Seismic Belt before the 2013 Lushan earthquake. Geodesy Geodynam 4(3):1–6. doi: 10.3724/SP.J.1246.2013.03001

Appendix 2
Forecasting the Strong Aftershock Hazard Following the May 12, 2008, Wenchuan Earthquake: A Personal Retrospective of the International Component

About 5 years after the May 12, 2008, Wenhcuan earthquake, a M_S7.0 earthquake occurred in Lushan, Sichuan Province, just in the southern part of the Longmenshan fault zone, in which the post-Wenchuan earthquake CFS modeling indicated an increased seismic hazard. The debate on whether this earthquake can be regarded as a 'late aftershock' of Wenchuan refreshes the retrospective investigation of the forecasting of the strong aftershocks of the Wenchuan sequence.

Estimate of the hazard of strong aftershocks, including the duration, place, maximum size, and expected rate, plays an important role in assisting the response to destructive earthquakes. Such an estimate has theoretical feasibilities due to the physical predictability of aftershock sequence.

After the May 12, 2008, Wenchuan earthquake, several experts from abroad were invited to participate in the discussion on the aftershock hazard. This Appendix summarizes the results obtained up to June 2008 and subject to test against the real aftershock activity till May 12, 2012. The materials might be interesting due to the real forward forecast test nature of this work, and marks the first time for Chinese seismological community to organize an international consultation targeting a specific requirement related to time-dependent seismic hazard.

A2.1 Estimate the Time-Dependent Hazard of Strong Aftershocks

Estimate of the (time-dependent) hazard of strong aftershocks, which deals with a limited spatial range and specific temporal duration, is a special usage of earthquake predictability with theoretical feasibility and practical implication (Tahir et al. 2012; Also see: Fig. A2.1). Questions addressed about this kind of forecast include the following issues:

Appendix 2 is written by Wu ZL.

Z. Wu et al., *Earthquake Phenomenology from the Field*, SpringerBriefs in Earth Sciences, 127
DOI: 10.1007/978-981-4585-15-6, © The Author(s) 2014

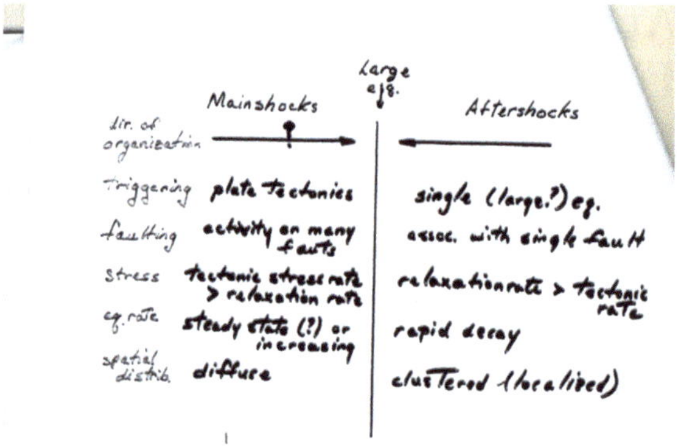

Fig. A2.1 Notes of Leon Knopoff (2007) (Lecture Notes of the Abdus Salam International Center for Theoretical Physics (ICTP), Trieste) showing the present understanding of aftershocks

Time—how long would be the duration of (strong) aftershocks;
Place—where would be the most-likely location of strong aftershocks;
Size—how big would be the maximum aftershock;
Rate—how many (strong) aftershocks would be expected in a given temporal duration, especially the period of rescue and relief.

After the May 12, 2008, Wenchuan earthquake in southwest China (Chen and Booth 2011), on May 21, the National Expert Committee on the Wenchuan Earthquake was established by the State Council. In the National Expert Committee, there were several working groups set, working on different urgent issues, from landslide disasters to reconstruction plans. I served as the vice chairman of the aftershock working group, with Guomin Zhang (China Earthquake Administration) being the chairman. On May 25, expressing gratitude to the colleagues of the IUGG Commission on Geophysical Risk and Sustainability (the GeoRisk Commission) for their kind words and offers to help, I wrote via e-mail:

> Now what I need urgently is the methods and results of the forecast of strong aftershocks which have been a great threaten to the relief and reconstruction process because in that region there have been quite a few landslide lakes. Any suggestion about this topic is welcome, and please contact me directly (and urgently) if there is any suggestion coming out from GeoRisk.

I noted specially: *Please note that when we discuss forecast we are meaning in a general sense. For example, how long will the aftershock sequence last, and what will be the magnitude of the largest aftershock.*

This request obtained active responses from colleagues within or not within the GeoRisk Commission. In this appendix I summarize the message feedback from the experts outside China. As a retrospective summary of the forecast endeavor, I focus on the 'predictive information' which means the estimate directly related to the time, location, size, and rate of the strong aftershocks. General comments on the Wenchuan earthquake and the physics of aftershocks are not included, except what R. Singh wrote on May 27, warning that:

> If I understood correctly, people living in the epicentral region are very much afraid and scared from the scientists/other people who are busy in giving false forecasts, during

> such critical time number of such people/groups increase significantly. We must resolve that such groups should not do anything, if they make such efforts they must inform the Government agencies, in case of China, one must inform CEA.

Actually all these comments are useful in the evaluation of the 'predictive information' and the decision-making for releasing the messages outside the scientific community.

A2.2 Estimate Based on Coulomb Failure Stress (CFS) Changes

As early as May 24, Jian Lin wrote:

> I am pretty certain that the Wenchuan quake has had important impact on the stress changes on faults in nearby regions. However, because this quake is of complex source mechanism rather than being a pure thrust fault, it is very important that we consider the source complexity carefully during the Coulomb stress modeling. Even for a pure thrust fault, it can induce large Coulomb stress changes on nearby strike-slip faults, as illustrated in Fig.4b (page 5) in the Lin and Stein (JGR 2004), which I am enclosing a copy and is also downloadable from the following site: http://quake.usgs.gov/research/deformation/modeling/papers/2004/Lin_Stein_JGR_2004.pdf.

Actually this is a 'trigger' of my message sent to the GeoRisk Commissioners. At the same time, study on the aftershock tendency was conducted by Chinese institutions, organized by the Department for Earthquake Monitoring and Forecast of the China Earthquake Administration (CEA), facilitated by the China Earthquake Networks Center (CENC). CEA formed a working group on aftershocks, concentrating on the works domestically conducted, and I also served as the vice chairman. The results obtained in China (Jiang et al. 2009) and those provided by colleagues abroad confirmed and complemented with each other. Logic for an international invitation was that since the aftershock problem is a major concern for the rescue and relief actions, and the Wenchuan earthquake has caused widespread attention around the world, let's have the worldwide top-ranked experts to discuss this problem—a problem with urgent and critical priority.

On May 28, Walter Mooney wrote to the Chinese Embassy to USA, with a copy to me:

> I would like to draw your attention to the two attached scientific papers written by several of the world's leading seismologists, including two from the US Geological Survey, Dr. Ross Stein and Dr. Tom Parsons.[1] These papers present calculations of the change in stress in the crust of western China following the May 12, 2008, Wenchuan earthquake.
>
> The paper by Toda and others, completed on May 22, 2008, correctly showed that there was increased stress (shown in RED on the color plot) at the northeast end of the fault. The destructive magnitude 6.0 aftershock occurred at this location on May 25, 2008, which seems to confirm the value of this calculation.

[1] Refer to the papers of Parsons et al. (2008) and Toda et al. (2008).

> There is also an increase in stress to the southwest of the fault. This means the faults in these regions are now more stressed.
>
> On the other hand, the main fault trace that moved on May 12, 2008, is now more relaxed, and has less stress.
>
> A similar important study has been completed by USGS seismologists Tom Parsons and his colleague Chen Ji, and I also forward that report to you. Please consult this paper to learn their main conclusions, as I will not repeat them all here.
>
> I have already forwarded both of these two important scientific papers to 50 scientific colleagues in China, including leading scientists at the CEA.
>
> My Chinese colleagues at the CEA are very knowledgeable, and are fully aware of these kinds of stress-change calculations. Perhaps they have already obtained similar results. However, as scientists, it is very useful to compare our results.

Indeed it is the case that Chinese seismologists also calculated the CFS change caused by the mainshock, which was published later.

A2.3 Analysis of Seismicity and Estimate of Aftershock Probability

A2.3.1 Analysis Based on the Gutenberg-Richter's Law and the Omori's Law

On May 25, Walter Mooney forwarded the message of Lucile Jones:

> This aftershock sequence, at least as recorded by USGS-NEIC, is well behaved. With two weeks of data and 97 aftershocks, the *b*-value appears stable with a value of 1.05. The sequence is decaying in a well-behaved way with a *p*-value of 0.97 ± 0.1. The *a*-value (as defined in Reasenberg and Jones 1989) is −2.15.
>
> Taking these parameters and turning them into rates and probabilities, we expect about 10 more aftershocks of at least *M*5.0 in the next 30 days. In the same time period (30 days starting from 12.5 days after the mainshock), the probability of at least one *M*6.0 is 70 %, the probability of a *M*6.5 or greater is 30 %, and the probability of a *M*7 or greater is 7 %.
>
> In general, this is an overall rather small aftershock sequence—for this size of mainshock. The "*a*-value" implies an overall rate 1/4 the average for California mainshocks. This *a*-value is the same we saw for Loma Prieta. Thus it seems likely there could be another magnitude 6, but a magnitude 7 is very unlikely. We believe (but do not yet have rigorous statistics to support) that a low *a*-value in the near-field aftershocks extends to the distant triggering as well and that would imply that the risk of a triggered earthquake in Xian or even farther east is unlikely.

On May 28, Mooney wrote:

> I spoke with Dr. Lucy Jones today at the USGS Menlo Park. She and Egill Hauksson have used your aftershock catalog to calculate time-dependent aftershock probabilities (Gerstenberger et al. Nature, 2005). Their results are consistent with their previous aftershock probability estimates, which I sent you about two days ago. In general, this aftershock sequence is relatively weak, with no aftershock that follows Bath's Law, that is, 1.2 magnitude units less than the mainshock (a magnitude 6.7 or 6.8 event). In their previous estimate, they gave a 7 % probability for a magnitude 7 aftershock (using the method of Reasenberg and Jones 1989).

A2.3.2 Estimating the Duration of the Strong Aftershock Sequence

On May 26, Harsh Gupta wrote:

> You have raised an extremely important question about how long the aftershocks would continue and what would be the largest aftershock's magnitude. I am not aware of any definite way of determining it. However, we can make some intelligent guesses by examining earlier earthquakes of similar size in the same region. Plotting of decay of aftershock activity of the past events and determining '*p*' value may be helpful.
>
> The 1950 *M* 8.7 India-China border earthquake (not very far away from the May 12, 2008 earthquake) had after shocks occurring for more than 2 years, and the largest aftershock was *M*~8! You may like to put some one to look at the past earthquakes, as you have a reasonably good catalogue of Chinese earthquakes. An important issue would be to re-examine earthquake magnitudes and the locations.

On June 23, Gupta wrote:

> Please refer to our discussions on the question as to how long the aftershocks of the Sichuan earthquake of May 12, 2008 would continue.
>
> We have carried out preliminary studies of the 'p' values from the available data of the aftershocks for the May 12 earthquake and some other recent and historical earthquake sequences. As $M >= 5$ aftershocks can be locally damaging, particularly in a region where the previous earthquakes have weakened the structures, we estimate that such aftershocks could occur for about 7 months after the main event, i.e. till January 2009. We have sent a short note on the same to the Geological Society of India.

A2.3.3 Analysis Based on the SSE Algorithm

On May 26, V. Kossobokov wrote:

> Eager to help with urgent analysis of what is going on in the region from the viewpoint of earthquake recurrences. You may recollect that pattern recognition studies by Inessa Vorobieva has led to so-called SSE algorithm dated back to 1992 (see the description http://www.mitp.ru/sse/SSE-Alg.html and early refs therein), which logic is in line with the Harsh's comment and suggestions.[2] The statistics of the on-going test of SSE confirm its high efficiency in predicting subsequent event of magnitude *M*-1 or larger (*M* being the magnitude of the earthquake in question). At the moment Inessa accumulates the data on the Wenchuan earthquake aftershocks required for a knowledgeable statement on the likelihood of subsequent strong earthquake (with magnitude 6.9 or larger) in the region.

On June 23, Kossobokov wrote:

> According to SSE algorithm applied to the aftershock sequence (I.A. Vorobieva, personal communication): Due to pattern recognition criteria NO MAGNITUDE 6.9 or larger event is expected to follow the 2008 Wenchuan (Sichuan, China) earthquake in the circle of 450 km radius centered at the epicenter of the 12 May 2008 main shock and in time

---

[2] See the last section of the comments of Harsh Gupta.

interval up to 10 Nov 2009. Note that in the forward testing mode to-date such a diagnosis made with SSE was confirmed in 13 out of 14 cases.

A2.3.4 ETAS Model-Based Analysis and Estimate of Aftershock Probabilities

ETAS model was used to characterize the aftershock sequence. Some estimate of the aftershock hazard was made based on the ETAS model. For example, on May 30, Jiancang Zhuang provided the analysis as follow:

Here are the probability forecasts for the 18th day after the mainshock.

	Expected #	Prob.	Waiting time	Quantile	(1 %	5 %	50 %	95 %	99 %)
M>=4.0	2.17	0.89	0.44		0.004	0.022	0.298	1.296	1.934
M>=4.5	1.03	0.64	0.89		0.010	0.048	0.605	2.655	4.061
M>=5.0	0.36	0.30	2.40		0.024	0.125	1.677	7.472	11.92
M>=5.5	0.12	0.11	6.53		0.058	0.339	4.306	20.08	32.89
M>=6.0	0.07	0.07	10.01		0.087	0.460	6.466	31.15	49.15

This analysis was lasted for some time. The whole aftershock zone is basically consistent with the rupture zone of the mainshock, which can be divided into two parts: the north part, and the south part. On June 5, by analyzing the aftershock activity, Zhuang detected a slow slip in the south, and estimated that *slow slip in the south may trigger the aftershocks in the north.*

A2.4 Real Situation of Aftershocks

Real situation of the Wenchuan aftershock sequence, with local network determined magnitude larger than 5.5, is shown in Table 5.1. The comparison of some of the descriptive estimates and the real situation is shown by Table A2.1. Table A2.2 lists the results independently obtained by Chinese seismological community.

Probabilities based on the calculated CFS changes, although having the Dieterich (1994) formula, are still hard to quantify and hard to be compared with real aftershock situation, which is still under study till present (Parsons et al. 2012). Therefore, the comparison in Table A2.1 does not include the CFS-related estimates. The same issue exists for the ETAS model-based estimate of the probabilities.

Table A2.1 mainly summarizes the analysis results based on the study of seismicity. It can be seen that, based on the present knowledge of aftershocks, it is possible to make some estimate of the maximum size, rate, duration, and location of strong aftershocks. How to interpret and use these pieces of information properly in the response to earthquakes is another important issue subject to discussion in future.

Table A2.1 Estimate of strong aftershock hazard with comparison with real situation

	Descriptive estimate and date	Real aftershock situation
1	There could be another magnitude 6, but a magnitude 7 is very unlikely (May 25)	There were 3 aftershocks over magnitude 6 occurred afterwards (2008/7/24 *M*6.0; 2008/8/01 *M*6.1; 2008/8/05 *M*6.1)
2	The risk of a triggered earthquake in Xian or even farther east is unlikely (May 25)	No earthquake larger than 5.5 occurred in Xi'an or farther east
3	No magnitude 6.9 or larger event is expected to follow the 2008 Wenchuan earthquake in the circle of 450 km radius centered at the epicenter of the 12 May 2008 main shock and in time interval up to 10 Nov 2009 (June 23)	Maximum magnitude of the aftershocks, after June, 2008, was 6.1
4	Slow slip in the south (identified by the ETAS-model fitting) may trigger the aftershocks in the north (June 5)	From July to September, 2008, all the aftershocks larger than 5.5 occurred in the north part of the aftershock zone
5	Such aftershocks could occur for about 7 months after the main event, i.e. till January 2009 (June 23)	From May 12 to December 31, 2008, 68 aftershocks over *M*5.0 occurred; From January 1, 2009 to December 31, 2011, 13 aftershocks over *M*5.0 occurred

Table A2.2 Analysis results of the Chinese institutions for comparison, with similar approaches to those mentioned in the text highlighted by italic, and with question marks indicating the cases that the evidences were not persuasive to me

Questions	Scientific basis	Estimates
Type of Sequence: mainshock-aftershock, or swarm?	*Analysis of the sequence parameters*	*Most likely mainshock-aftershock*
	Earthquakes in continental China: 18 % swarms	Most likely (?) mainshock-aftershock
	Earthquakes over *M*7.8 in continental China: 2 out of 10 swarms	Most likely (?) mainshock-aftershock
	Global thrust earthquakes over *M*7.8: ~18 % swarms	Most likely (?) mainshock-aftershock
	Earthquake history along the Longmenshan fault: all mainshock-aftershock	Most likely mainshock-aftershock
	log Δt – log t relation	Most likely mainshock-aftershock
Time length of the sequence	*Analysis of the sequence parameters*	*~150 days for M6 aftershocks*
	Aftershock sequences in China: 95 % maximum aftershocks occurred within 150 days	~5 months for strong aftershocks
	Statistics of aftershock (over *M*6) sequences in China	~2 months for *M*6 aftershocks
	Statistics of aftershock (over *M*5) sequences in China	Over 1 year for *M*5 aftershocks

(continued)

Table A2.2 (continued)

Questions	Scientific basis	Estimates
	Statistics of aftershock sequences in Sichuan-Yunnan	~2 months for *M*6 aftershocks
	Statistics of aftershock sequences of global thrust earthquakes over *M*7.5: ~95% maximum aftershocks occurred within 40 days	~40 days for *M*6 aftershocks
Maximum magnitude of aftershocks	*High seismic wave radiation efficiency or high stress drop*	*Mainshock-aftershock magnitude difference > 1.0*
	Size of aftershock zone	Maximum 6.2–6.6
	Size of surface fracture	Less than 7
	Bath's law	~6.8 but with uncertainty (?)
	Bath's law for thrust events	~6.6
	GR-law of aftershock sequence and the Zipf rank-ordering extrapolation	*~6.7*
	Earthquakes over 7.5 in the north-south seismic zone: 21 out of 23 with aftershocks less than 7	Less than 7
	Global thrust earthquakes over 7.5	~70 % probability less than 7 (?)
	Rupture process and the 'gap' or 'deficit' along the rupture fault (?)	~*M*7 or even larger (?)
Location of strong aftershocks: the north part, or the south part?	Higher apparent stress in the north	Most likely the north part
	Northward 'migration' (?) of aftershocks over 5	Most likely the north part

Note Results were mainly from: the Institute of Geophysics, CEA; the Institute of Earthquake Science, CEA, the Institute of Crustal Geodynamics, CEA; the China Earthquake Networks Center (CENC); the Earthquake Administration of Sichuan Province; and the Earthquake Administration of Hebei Province, with the contribution from the University of Science and Technology of China (USTC), the Institute of Geology and Geophysics of the Chinese Academy of Sciences (CAS), and the Institute of Geodesy and Geophysics of CAS (Jiang et al. 2009). Temporal forecasts were also tried and tested (for example, see, Jiang and Wu, 2012) but are not discussed here

A2.5 Concluding Remarks

A2.5.1 Conclusions and Discussion

This Appendix summarizes the estimate of strong aftershock hazard following the May 12, 2008, Wenchuan, southwest China, earthquake, provided by colleagues abroad. Three issues are remarkable for this endeavor. (1) The estimate of the aftershock hazard provides an important case to make full use of the predictability of earthquakes, which can help much to the response to destructive earthquakes. (2) In the test of earthquake forecast, real forward forecast plays a crucial role (Mulargia 1997). The Wenchuan case provided a good example of real forward forecast test. The retrospective comparison with real aftershock situation shows that the predictive estimate provided *a priori* are basically in consistence with real seismicity. (3) Earthquake science, as well as its application to the reduction of seismic disasters, needs international collaboration. This work was the first time for Chinese seismologists to have an international 'Dizhen Huishang' (Predictive Consultation on the Likelihood of Earthquakes) facing to real earthquake situation, although as early as the 1990s Chinese seismological agencies invited some foreign scientists attending the Annual Consultation on the Likelihood of Earthquakes (Wu 1997). While in history, there were quite a few lessons about the cross-border forecast of earthquake hazard, the Wenchuan case showed that the invitation-based consultation, as a complement to the domestic studies underway, may be of help to the rescue and relief actions, without negative social effects. I believe that this case is interesting not only in the history of seismology but also in the study of earthquake hazard.

A2.5.2 Additional Remarks

In 2008, during the Sino-US Joint Workshop on Seismological Studies (Boulder, Colorado), I presented a poster describing the forecast of the aftershocks of the Wenchuan earthquake. Before that, I circulated within the GeoRisk Commission an open file describing this work, and got positive and constructive feedback from the colleagues related. I am grateful for all of the experts whose e-mail communications were permitted to publish.

Besides the colleagues mentioned in the text, I would like to thank G. F. Panza who gave useful comments (on May 13) on the geodynamics of the Wenchuan earthquake, Peter Bormann who provided the fast estimate of the energy release of the Wenchuan mainshock (on May 22) indicating a high stress drop, E. R. Enghahl and E. Bergman who provided (on May 23) the EHB re-location of the Wenchuan aftershock sequence, N. Hirata who provided the ETAS analysis result of the Wenchuan aftershock series (on June 4), Y. Kontar who gave useful comments on the properties and physics of aftershocks (replying my e-mail on May 25), X. D. Song (on May 28) who provided important references[3] for the aftershock analysis, J. Derr who provided the information about the reported pre-shock anomaly (on May 23), Alik Ismail-Zadeh who facilitated the discussion within the GeoRisk Commission to a considerable extent, and late Professor L. Knopoff who gave useful comments (on June 6) on the pre-shock seismicity. All these messages were helpful in the seismological response to the Wenchuan earthquake.

I also take this opportunity to thank H. Gupta, V. Kossobokov, G. F. Panza, D. Rhoades, J. Rundle, A. Zavyalov, and C. Z. Zhu who helped in the review of an earthquake forecast for summer 2008 in north China. Although this forecast

[3] References provided were Reasenberg and Jones (1989, 1994).

(published in 2007) was not directly related to the aftershock sequence or the 'remote triggering' effect of the Wenchuan earthquake, it caused attention in China after the Wenchuan earthquake, with the approaching of the Beijing Olympic Games. This review, conducted in May, 2008, after the Wenchuan earthquake, can be understood to a considerable extent as the 'after-effect' of the forecast work mentioned here, although the two works were almost parallel to each other. The expert review rejected the forecast, and the review conclusion was shown to be in consistence with real earthquake situation.

Acknowledgments
Thanks to Dongning Zhang and Changsheng Jiang, Institute of Geophysics, CEA, and Li Li, National Earthquake Infrastructures Center, CEA, for assistance in the related works. I thank Shi Che, deputy of the Department for Earthquake Monitoring and Forecast, CEA, and Guomin Zhang, chairman of the aftershock working group of the National Expert Committee, for authorization and endorsement of the organization of this 'international consultation'.

References

Chen Y, Booth DC (2011) The Wenchuan earthquake of 2008: anatomy of a disaster. Science Press in cooperation with Springer, Beijing

Dieterich JH (1994) A constitutive law for rate of earthquake production and its application to earthquake clustering. J Geophys Res 99:2601–2618

Gerstenberger MC, Wiemer S, Jones LM, Reasenberg PA (2005) Real-time forecasts of tomorrow's earthquakes in California. Nature 435:328–331

Jiang CS, Wu ZL (2012) Testing the forecast of aftershocks: a simple method with an example of application. Res Geophys 2(e5):29–33. doi:10.4081/rg.2012.e5

Jiang CS, Zheng Y, Zhou LQ (2009) Aftershock sequence. In: Science report on the Wenchuan Magnitue 8.0 earthquake. Seismological Press, Beijing, pp 130–169 (in Chinese)

Mulargia F (1997) Retrospective validation of the time association of precursors. Geophys J Int 131:500–504

Parsons T, Ji C, Kirby E (2008) Stress changes from the 2008 Wenchuan earthquake and increased hazard in the Sichuan basin. Nature 454:509–510

Parsons T, Ogata Y, Zhuang J, Geist EL (2012) Evaluation of static stress change forecasting with prospective and blind tests. Geophys J Int 188:1425–1440. doi:10.1111/j.1365-246X.2011.05343.x

Reasenberg PA, Jones LM (1989) Earthquake hazard after a mainshock in California. Science 243:1173–1176

Reasenberg PA, Jones LM (1994) Earthquake aftershocks: update. Science 265:1251–1252

Tahir M, Grasso J-R, Amorèse D (2012) The largest aftershock: how strong, how far away, how delayed? Geophys Res Lett 39(4):L04301. doi:10.1029/2011GL050604

Toda S, Lin J, Meghraoui M, Stein RS (2008) 12 May 2008 M = 7.9 Wenchuan, China, earthquake calculated to increase failure stress and seismicity rate on three major fault systems. Geophys Res Lett 35:L17305, doi:10.1029/ 2008GL034903

Wu FT (1997) The annual earthquake prediction conference in China (National consultative meeting on seismic tendency). Pure appl Geophys 149:249–264

MIX
Papier aus verantwortungsvollen Quellen
Paper from responsible sources
FSC® C105338

If you have any concerns about our products,
you can contact us on
ProductSafety@springernature.com

In case Publisher is established outside the EU,
the EU authorized representative is:
Springer Nature Customer Service Center GmbH
Europaplatz 3, 69115 Heidelberg, Germany

Printed by Libri Plureos GmbH
in Hamburg, Germany